SYSTEM-LEVEL DESIGN TECHNIQUES FOR ENERGY-EFFICIENT EMBEDDED SYSTEMS

T0135119

System-Level Design Techniques for Energy-Efficient Embedded Systems

by

MARCUS T. SCHMITZ
University of Southampton, United Kingdom

BASHIR M. AL-HASHIMI
University of Southampton, United Kingdom

and

PETRU ELES
Linköping University, Sweden

KLUWER ACADEMIC PUBLISHERS
BOSTON / DORDRECHT / LONDON

Published by Kluwer Academic Publishers,
P.O. Box 17, 3300 AA Dordrecht, The Netherlands.

Sold and distributed in North, Central and South America
by Kluwer Academic Publishers,
101 Philip Drive, Norwell, MA 02061, U.S.A.

In all other countries, sold and distributed
by Kluwer Academic Publishers,
P.O. Box 322, 3300 AH Dordrecht, The Netherlands.

Printed on acid-free paper

ISBN 978-1-4419-5429-9   e-ISBN 978-0-306-48736-1

Printed in the Netherlands.

To our beloved families

Contents

List of Figures

List of Tables

Preface

It is likely that the demand for embedded computing systems with low energy dissipation will continue to increase. This book is concerned with the development and validation of techniques that allow an effective automated design of energy-efficient embedded systems. Special emphasis is placed upon system-level co-synthesis techniques for systems that contain dynamic voltage scalable processors which can trade off between performance and power consumption during run-time.

The first part of the book addresses energy minimisation of distributed embedded systems through dynamic voltage scaling (DVS). A new voltage selection technique for single-mode systems based on a novel energy-gradient scaling strategy is presented. This technique exploits system idle and slack time to reduce the power consumption, taking into account the individual task power dissipation. Numerous benchmark experiments validate the quality of the proposed technique in terms of energy reduction and computational complexity.

The second part of the book focuses on the development of genetic algorithm-based co-synthesis techniques (mapping and scheduling) for single-mode systems that have been specifically developed for an effective utilisation of the voltage scaling approach introduced in the first part. The schedule optimisation improves the execution order of system activities not only towards performance, but also towards a high exploitation of voltage scaling to achieve energy savings. The mapping optimisation targets the distribution of system activities across the system components to further improve the utilisation of DVS, while satisfying hardware area constraints. Extensive experiments including a real-life optical flow detection algorithm are conducted, and it is shown that the proposed co-synthesis techniques can lead to high energy savings with moderate computational overhead.

The third part of this book concentrates on energy minimisation of emerging distributed embedded systems that accommodate several different applications within a single device, i.e., multi-mode embedded systems. A new co-synthesis technique for multi-mode embedded systems based on a novel operational-mode-state-machine specification is presented. The technique increases significantly the energy savings by considering the mode execution probabilities that yields better resource sharing opportunities.

The fourth part of the book addresses dynamic voltage scaling in the context of applications that expose extensive control flow. These applications are modelled through conditional task graphs that capture control flow as well as data flow. A quasi static scheduling technique is introduced, which guarantees the fulfilment of imposed deadlines, while at the same time, reduces the energy dissipation of the system through dynamic voltage scaling.

The new co-synthesis and voltage scaling techniques have been incorporated into the prototype co-synthesis tool LOPOCOS (Low Power Co-Synthesis). The capability of LOPOCOS in efficiently exploring the architectural design space is demonstrated through a system-level design of a realistic smart phone example that integrates a GSM cellular phone transcoder, an MP3 decoder, as well as a JPEG image encoder and decoder.

Acknowledgments

Financial support of the work was provided by the Department of Electronics and Computer Science at the University of Southampton, the Embedded Systems Laboratory (ESLAB) at Linköping University, as well as the Engineering and Physical Sciences Research Council (EPSRC), UK.

Special thanks go to the members of the Electronic Systems Design Group (ESD) at the University of Southampton, for many fruitful discussions.

We would like to thank Christian Schmitz, who has contributed in deriving the smart phone benchmark during a visit at the University of Southampton.

We would also like to acknowledge Neal K. Bambha (University of Maryland, USA) and Flavius Gruian (Lund University, Sweden) for kindly providing their benchmark sets, which have been used to conduct some of the presented experimental results.

Chapter 1

INTRODUCTION

Over the last several years, the popularity of portable applications has explosively increased. Millions of people use battery-powered mobile phones, digital cameras, MP3 players, and personal digital assistants (PDAs). To perform major parts of the system's functionality, these mass products rely, to a great extent, on sophisticated embedded computing systems with *high performance* and *low power dissipation*. The complexity of such devices, caused by an ever-increasing demand for functionality and feature richness, has made the design of modern embedded systems a time-consuming and error-prone task. To be commercially successful in a highly competitive market segment with tight time-to-market and cost constraints, computer-based systems in mobile applications should be cheap and quick to realise, while, at the same time, consume only a small amount of electrical power, in order to extend the battery-lifetime. Designing such embedded systems is a challenging task.

This book addresses this problem by providing techniques and algorithms for the automated design of energy-efficient distributed embedded systems which have the potential to overcome traditional design techniques that neglect important *energy management* issues. In this context, special attention is drawn to *dynamic voltage scaling* (DVS) — an energy management technique. The main idea behind DVS is to dynamically scale the supply voltage and operational frequency of digital circuits during run-time, in accordance to the temporal performance requirements of the application. Thereby, the energy dissipation of the circuit can be reduced by adjusting the system performance to an appropriate level. Furthermore, the proposed synthesis techniques target the coordinated design (co-design) of *mixed hardware/software* applications towards the effective exploitation of DVS, in order to achieve substantial reductions in energy.

The main aims of this chapter are to introduce the fundamental problems that are involved in designing distributed embedded systems and to provide

1

the terminology used throughout this work. The remainder of this chapter is organised as follows. Section 1.1 outlines a typical system-level design process. A task graph specification model, used to capture the system's functionality, is introduced in Section 1.2. Section 1.3 describes the individual system design steps using some illustrative examples. Hardware and software synthesis are briefly discussed in Section 1.4. Finally, Section 1.5 gives an overview of the book contents.

1.1 Embedded System Design Flow

A typical embedded system, as it can be found, for example, in a smart-phone, is shown in Figure 1.1. It consists of heterogeneous components such

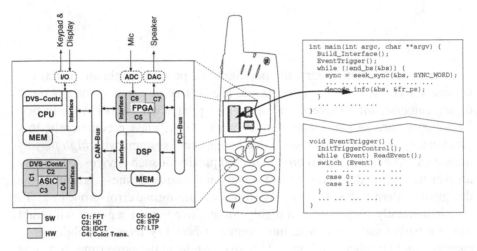

(a) Embedded architecture: A distributed heterogeneous system

(b) Embedded software

Figure 1.1.   Example of a typical embedded system (smart-phone)

as software programmable processors (CPUs, DSPs) and hardware blocks (FP-GAs, ASICs). These components are interconnected through communication links and form a distributed architecture, such as the one shown in Figure 1.1(a). Analogue-to-digital converters (ADC), digital-to-analogue converters (DAC), as well as input/output ports (I/O) allow the interaction with the environment. A complete embedded system, however, consists additionally of application software (Figure 1.1(b)) that is executed on the underlying hardware architecture (Figure 1.1(a)). Clearly, effective embedded system design demands optimisation in *both* hardware and software parts of the application. When designing an embedded computing system, as part of a new product, it is common to go through several design steps that bring a novel product idea down to its physical realisation. This is usually referred to as system-level design flow. A possible and common design flow is introduced in Figure 1.2. It is characterised by three important design steps: *system specification* (Step A), *co-synthesis* (Step B), as well as concurrent *hardware and software synthesis* (Step C). The remainder of this section briefly outlines this design flow.

Starting from a new product idea, the first step towards a final realisation is *system specification*. At this stage, the functionality of the system is captured using different conceptual models [61] such as natural language, annotated-graphic representations (finite state machines, data-flow graphs), or high-level languages (VHDL, C/C++, SystemC). This design step is indicated as Step A in Figure 1.2. Having specified the system's functionality, the next stage in the design flow is the *co-synthesis*, shown as Step B in Figure 1.2. The goal of co-synthesis is threefold:

Architecture allocation: Firstly, an adequate target architecture needs to be allocated, i.e., it is necessary to determine the quantity and the types of different interconnected components that form the distributed embedded system. Components that can be allocated are given in a predefined technology library.

Application mapping: Secondly, all parts of the system specification have to be distributed among the allocated components, that is, tasks (function fragments) and communications (data transfers between tasks) are uniquely mapped to processing elements and communication links, respectively.

Activity scheduling: Thirdly, a correct execution order of tasks and communications has to be determined, i.e., the activities have to be scheduled under the consideration of interdependencies.

These three co-synthesis stages aim to optimise the design according to objectives set by the designer, such as power consumption, performance, and cost. In order to reduce the power consumption, emerging co-synthesis approaches (as the one proposed in this work) tightly integrate the consideration of *energy management techniques* within the design process [67, 76, 99, 100].

Energy management Energy management techniques utilise existing idle times to reduce the power consumption by either shutting down the idle components or by reducing the performance of the components.

The consideration of energy management techniques during the co-synthesis allows the optimisation of allocation, mapping, and scheduling towards their effective exploitation. After the co-synthesis has allocated an architecture as well as mapped and scheduled the system activities (tasks and communications), the next stage in the design flow is the concurrent *hardware and software synthesis*, indicated as Step C in Figure 1.2. These separated design steps transform the system specification, which has been split between hardware and software, into

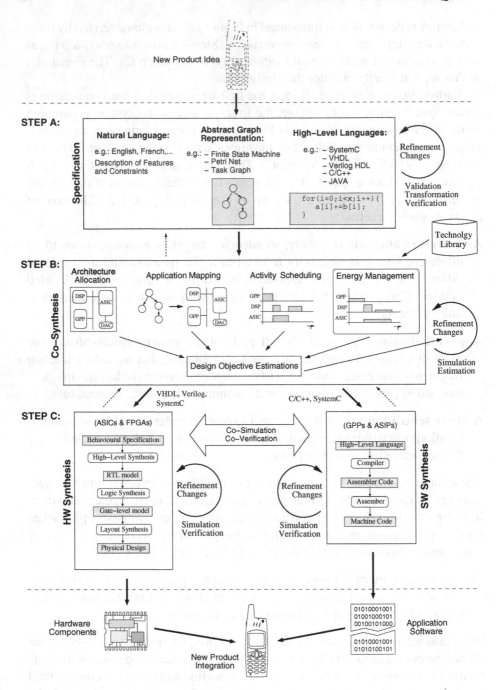

Figure 1.2.   Typical design flow of a new embedded computing system

physical implementations. System parts that are mapped onto customised hardware are designed using high-level [8, 19, 60, 134, 154], logic [9, 42, 110, 131], and layout [56] synthesis tools. While system parts that have been mapped onto software programmable processors (CPUs, DSPs) are compiled into assembler and machine code, using either standard or specialised compilers and assemblers [1, 93]. The main advantage of a concurrent hardware (HW) and software (SW) synthesis is the possibility to co-simulate both system parts, with the aim of finding errors in the design as early as possible to avoid expensive re-designs. The following section describes the whole design process shown in Figure 1.2 in more detail and introduces the terminology used throughout this book.

1.2 System Specification (Step A)

The functionality of a system can be captured using a variety of conceptual specification models [61]. Different modelling styles are, for example, high-level languages (hardware description and programming languages) such as SystemC, Verilog HDL, VHDL, C/C++, or JAVA, as well as more abstract models such as block diagrams, task graphs, finite state machines (FSMs), Petri nets, or control/dataflow graphs. Typical applications targeted by the presented work can be found in the audio and video processing domain (e.g. multi-media and communication devices with extensive data stream operations). Such applications fall into the category of data-flow dominated systems. An appropriate representation for these systems is the task graph model [84, 112, 157], which will be introduced in the following section.

1.2.1 Task Graph Representation

The functionality of a complex system with intensive data stream operations can be abstracted as a directed, acyclic graph (DAG) $G_S = (\mathcal{T}, \mathcal{C})$, where the set of nodes $\mathcal{T} = \{\tau_0, \tau_1, ..., \tau_n\}$ denotes the set of tasks to be executed, and the set of directed edges $\mathcal{C}$ refers to communications between tasks, with $\gamma_{ij} \in \mathcal{C}$ indicating a communication from task τ_i to task τ_j. A task can only start its execution after all its ingoing communications have finished. Each task can be annotated with a deadline θ, the time by which its execution has to be finished. Furthermore, the task graph inherits a repetition period ϕ which specifies the maximal delay between to invocations of the source tasks (tasks with no ingoing edges). Structurally, task graphs are similar to the data-flow graphs that are commonly used in high-level synthesis [60, 154]. However, while nodes in data-flow graphs represent single operations, such as multiplications and additions, the nodes in task graphs are associated with larger (coarse) fragments of functionality, such as whole functions and processes. The concept behind this model can be exemplified using a simple illustrative example.

Example 1:   For the purpose of this example, consider an MP3 audio decoder. In order to reconstruct the "original" stereo audio signal from an encoded stream, the decoder reads the data stream and applies several transformations such as Huffman decoding, dequantisation, inverse discrete cosine transformation (IDCT), and antialiasing. A possible task graph specification along with a high-level language description in C of such an MP3 decoder is shown in Figure 1.3. The figure outlines the relation between task graph model and high-level description. In this particular example the granularity of each task in the task graph corresponds to a single sub-function of the C specification. For instance, the Huffman Decoder tasks (τ_3 and τ_4) in Figure 1.3(a) reflect the functionality that is performed by the third sub-function in Figure 1.3(b). The flow of data is expressed by edges between the individual tasks. The output data produced by the Huffman Decoder tasks, for example, is the input of the dequant tasks (τ_5 and τ_6), indicated by the communication edges $\gamma_{3,5}$ and $\gamma_{4,6}$. In order to decode the compressed data into a high quality audio signal, one execution of all tasks in the graph, starting from task τ_0 and finishing with τ_{16}, has to be performed in at most $25ms$ as expressed by the task deadline θ_{16}. However, to obtain real-time decompression of a continuous music stream, the execution of all tasks has to be performed 40 times per second, i.e., with a repetition rate of $\phi = 25ms$. Although in this particular example the deadline and the repetition rate are identical, they might vary in other applications. As opposed to the C specification, the task graph explicitly exhibits *application parallelism* as well as *communication between tasks (data flow)*, while the exact algorithmic implementation of each function is abstracted away.   □

Task graphs can be derived from given high-level specification either manually or using extraction tools, such as the one proposed in [148].

1.3   Co-Synthesis (Step B)

Once the system's functionality has been specified as task graph, the system designers will start with the system-level co-synthesis. This is indicated as Step B in Figure 1.2. In addition, Figure 1.4 shows the co-synthesis flow in diagrammatic form. Co-synthesis is the process of deriving a mixed hardware/software implementation from an abstract functional specification of an embedded system. To achieve this goal, the co-synthesis needs to address four fundamental design problems: *architecture allocation*, *application mapping*, *activity scheduling*, and *energy management*. Figure 1.4 shows the order in which these problems have to be solved. In general, these co-synthesis steps are iteratively repeated until all design constraints and objectives are satisfied [52, 54, 70, 156]. An iterative design process has the advantage that valuable feedback can be provided to the different synthesis steps. This feedback, which

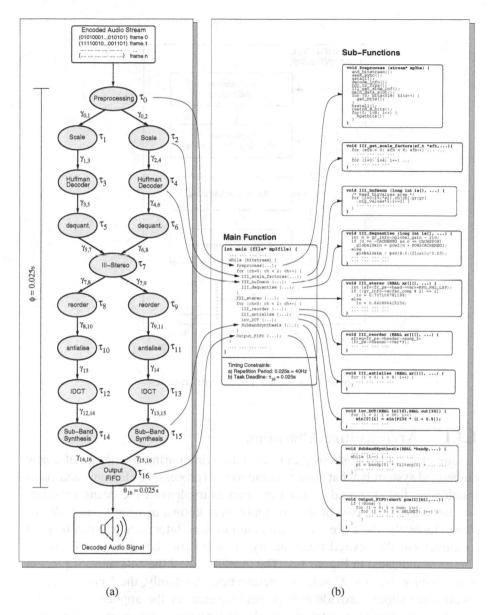

Figure 1.3. MP3 decoder given as (a) task graph specification (17 tasks and 18 communications) and (b) high-level language description in C

is indicated by dashed upwards arrows in the figure, is used to guide the optimisation process towards the satisfaction of design constraints. The following sections explain the co-synthesis flow shown in Figure 1.4 and the four subproblems in more detail.

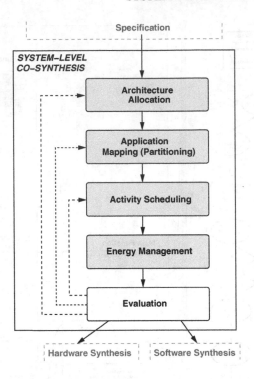

Figure 1.4.   System-level co-synthesis flow

1.3.1   Architecture Allocation

One of the first questions that needs answering during the design of a new embedded system is what system components (processing elements and communication links) should be used in order to implement the desired product functionality. This part of the co-synthesis is known as *architecture allocation*. Generally, there are many different target architectures that can be used to implement the desired functionality. Problematic, however, is the correct choice as indicated in Figure 1.5. The overall goal of the co-synthesis process is to identify the "most" suitable architecture. Certainly, the "most" suitable architecture should provide enough performance for the application in order to satisfy the timing constraints, while, at the same time, cost, design time, and energy dissipation should be reduced to a minimum. The importance of architecture allocation becomes clearer when considering the advantages and disadvantages associated with processing elements of various kinds. Table 1.1 gives the most relevant component trade-offs. Consider, for instance, the two processing elements (PEs): general-purpose processor (GPP) and application specific integrated circuit (ASIC). While software implementations on off-the-self GPPs are more flexible and cheaper to realise than hardware designs, the

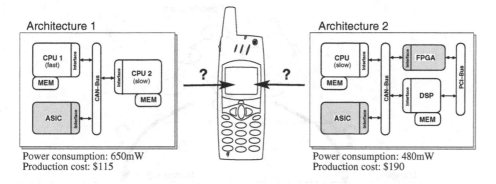

Architecture 1   Architecture 2

Power consumption: 650mW
Production cost: $115

Power consumption: 480mW
Production cost: $190

Figure 1.5. Architectural selection problem

	GPPs	ASIPs	FPGAs	ASICs
Cost	+ +	+ +	o	- -
Flexibility	+ +	+	o	- -
Performance	-	o	+	+ +
Energy-Efficiency	-	o	+	+ +

Table 1.1. Trade-offs between serveral heterogeneous components
(+ + ≡ highly advantageous, + ≡ advantageous, o ≡ moderate,
- ≡ disadvantageous, - - ≡ highly disadvantageous)

ASIC offers higher performance and better energy-efficiency. Similarly, the application specific instruction set processors (ASIPs) and field-programmable gate arrays (FPGAs) show different trade-offs. Of course, the non-recurring engineering cost (NRE) is mainly important for low volume products. For high volume applications this cost is amortised and becomes less important. Certainly, selecting the appropriate system components, in order to balance between these trade-offs, is of utmost importance for high quality designs. The intention of system-level co-synthesis tools is to aid the system designer in effectively exploring the architectural design space, in order to find a suitable target architecture rapidly.

1.3.2 Application Mapping

Following the co-synthesis flow given in Figure 1.4, the next step after architecture allocation is *application mapping*. During this step the tasks and communications of the system specification are mapped onto the allocated processing elements (PEs) and communication links (CLs) of the architecture, respectively. Figure 1.6 illustrates two different mappings of a system specification onto identical target architectures. These two mappings differ in the assignment of task τ_4, which is either mapped to the ASIC (Mapping 1) or to

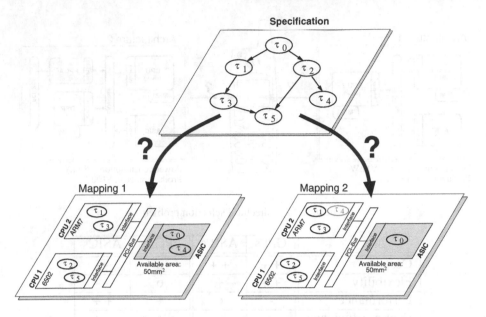

Figure 1.6. Application mapping onto hardware and software components

CPU2 (Mapping 2). Mapping explicitly determines if a task is implemented in hardware or software, hence, the term *hardware/software partitioning* is often mentioned in this context. Due to the heterogeneity of processing elements, the mapping specifies the execution characteristics of each task and communication. Consider, for example, the execution characteristics of the tasks shown in Table 1.2. This table gives the execution times t_{exe} and power dissipations

| | CPU 1 | | CPU 2 | | ASIC (50mm^2, | | |
| | (6502 @10MHz) | | (ARM7 @20MHz) | | Technology: 0.6μm) | | |
Task	t_{exe} (ms)	P_{dyn} (mW)	t_{exe} (ms)	P_{dyn} (mW)	t_{exe} (ms)	P_{dyn} (mW)	A (mm^2)
τ_0	89.3	3.6	12.1	23	1.8	0.13	7.76
τ_1	25.2	3.9	3.0	26	0.3	0.05	5.82
τ_2	19.7	4.4	2.8	28	0.2	0.07	9.71
τ_3	31.1	3.8	4.7	28	0.4	0.02	12.52
τ_4	172.2	3.9	22.3	27	2.7	0.12	8.05
τ_5	27.2	4.2	3.5	24	0.6	0.02	3.74

Table 1.2. Task execution properties (time and power) on different processing elements

P_{dyn} of each task in the specification of Figure 1.6, depending on the mapping to a 6052 8-bit microprocessor (running at 10MHz), an ARM7TDMI

32-bit microprocessor (running at $20MHz$), or an ASIC in $0.6\mu m$ technology which offers a usable die size of $50mm^2$. In addition to the time and power values, the hardware area A required for tasks implemented on the ASIC is given. In general, hardware implementations are more efficient in terms of performance and power consumption than software realisations. However, the design of hardware is a more time consuming process. Clearly, determining a good mapping solution is of crucial importance for the system design. Inappropriately distributing the activities among the components can result in poor utilisation of the system, necessitating the allocation of an architecture with higher performance, hence, increasing the system cost.

1.3.3 Activity Scheduling

Moving further in the design flow of Figure 1.4, the next step after application mapping is *activity scheduling*. The function of scheduling is to order the execution of tasks and communications (both activities) such that timing constraints are satisfied. This is not a trivial problem, since several activities mapped onto the same component cause congestion, which, in turn, hampers the effective exploitation of parallelism in the application. Hence, a good schedule should allow to exploit this parallelism effectively in order to improve the system performance.

Given an allocated architecture and a mapping of tasks and communication as well as a task graph specification, Figure 1.7 depicts two possible schedule solutions (Schedule 1 and Schedule 2). According to the system specification, the execution of the tasks τ_4 and τ_5 must be finished before deadline θ is exceeded. Thus, if the deadlines are violated the schedule is invalid. Consider the following scheduling scenarios given in Schedule 1 and Schedule 2 of Figure 1.7. After the initial task τ_0 has finished its execution, the communications $\gamma_{0,1}$ and $\gamma_{0,2}$ become ready. However, since both communications need to share the same bus it is necessary to sequence the transfers, since only one transfer is possible at a given time. Thus, a scheduling decision has to be taken at this point. The first schedule shown in Figure 1.7 corresponds to a schedule in which communication $\gamma_{0,1}$ takes place before communication $\gamma_{0,2}$. As it can be observed from this schedule, the executions of tasks τ_4 and τ_5 finish before deadline θ, hence, this solution represents a valid schedule ($t_{f1} < \theta$). On the other hand, if communication $\gamma_{0,2}$ is scheduled before communication $\gamma_{0,1}$, as shown in Schedule 2, the execution of task τ_1 is delayed, which further delays task τ_3 and communication $\gamma_{3,5}$. Ultimately, the execution of task τ_5 starts too late to finish the execution before deadline θ. Thus, the second schedule represents an invalid solution ($t_{f2} > \theta$).

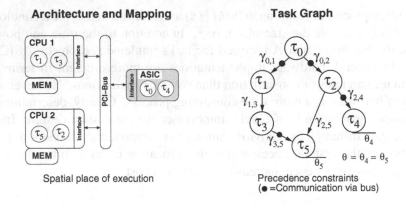

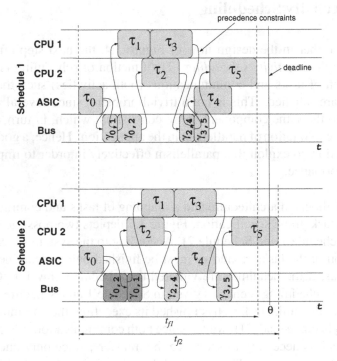

Figure 1.7. Two different scheduling variants based on the same allocated architecture and identical application mapping

1.3.4 Energy Management

Having allocated an architecture as well as having mapped and scheduled the application onto it, the next step within the co-synthesis flow of Figure 1.4 is the utilisation of *energy management techniques*. This step is necessary to accurately estimate the energy requirements of the system, which is used to guide the optimisation of allocation, mapping, and scheduling towards energy-

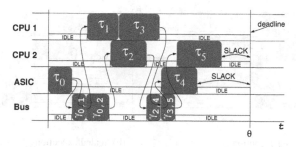

Figure 1.8. System schedule with idle and slack times

efficient designs. In general, energy management techniques exploit idle times and slack times within the system schedule by shutting down processing elements (PEs) [26, 97] or by reducing the performance of individual PEs [36, 152]. Idle times and slack times are defined as follows:

- **Idle times** refer to periods in the schedule when PEs and CLs do not experience any workload, i.e., during these intervals the components are redundant (see Figure 1.8).

- **Slack times** is the difference between task deadline and task finishing time of sink tasks (tasks with no outgoing edges), i.e., slack times are a result of over-performance (see Figure 1.8). Clearly, slack time is a special case of idle time.

Two important energy management techniques are dynamic power management (DPM) [26, 79, 97, 140] and dynamic voltage scaling (DVS) [36, 76, 80, 152]. DPM puts processing elements and communication links (both components) into standby or sleeping modes whenever they are idle. Nevertheless, the reactivation of components takes finite time and energy; hence, components should only be switched off or set into a standby mode if the idle periods are long enough to avoid deadline violations or increased power consumptions [26, 98]. DVS, on the other hand, exploits slack time by reducing simultaneously clock frequency and supply voltage of PEs. Thereby, DVS adapts the component performance to the actual requirement of the system. In this way, substantial savings are achieved since the energy consumption of the system components is proportional to the square of the supply voltage ($E \propto V^2$) [38]. The basic concept behind DVS is demonstrated in Figure 1.9. It can be observed from Figure 1.9(a) that tasks τ_4 and τ_5 finish execution before deadline θ. As indicated in the figure, this results in slack time. Instead of switching-off the components during these times (as done by DPM), it is possible to prolong the execution of all six tasks. This is achieved by scaling down the supply voltage and frequency of the processing elements until the tasks τ_4 and τ_5 just finish on time (as shown in Figure 1.9(b)). The main problem that needs to be addressed

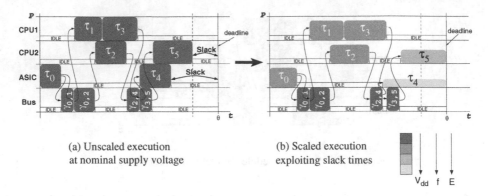

(a) Unscaled execution   (b) Scaled execution
at nominal supply voltage   exploiting slack times

Figure 1.9.   The concept of dynamic voltage scaling

here is how to distribute the available slack time among the tasks, in order to achieve the "highest" possible energy savings.

Nevertheless, the effectiveness with which DPM and DVS can be applied depends significantly on the available idle and slack times. A worthwhile optimisation of allocation, mapping, and scheduling must take the optimisation of idle and slack time into account, in order to allow a most effective exploitation of both techniques [68, 76, 98, 99]. In general, such an optimisation requires the iterative execution of the co-synthesis steps (allocation, mapping, scheduling), until the "most" suitable implementation of the system has been found [67, 99].

1.4   Hardware and Software Synthesis (Step C)

The previous section has outlined the system-level co-synthesis (Step B in Figure 1.2), which transforms an abstract specification into an architectural description of a mixed hardware/software system. The final step in the embedded system design flow is the concurrent *hardware and software synthesis* (Step C in Figure 1.2). This step brings the mixed hardware/software description of the system down to a physical implementation, i.e., the specification fragments (tasks) that have been distributed among the hardware and software components of the system need to be realised. This is achieved through two separate, yet concurrent synthesis steps: hardware synthesis and software synthesis. One of the main advantages of concurrent HW/SW design is the ability to check the correctness of the overall system by means of simulation, i.e., the interaction between hardware and software can be co-simulated [129]. Note, whereas system-level co-synthesis targets the design of interacting components, the main aim of hardware and software synthesis is the design of the *individual* hardware components and the software tasks running on programmable processors.

Hardware Synthesis: The design of complex hardware components is based on existing very large scale integration (VLSI) synthesis tools [8, 9, 43, 134,

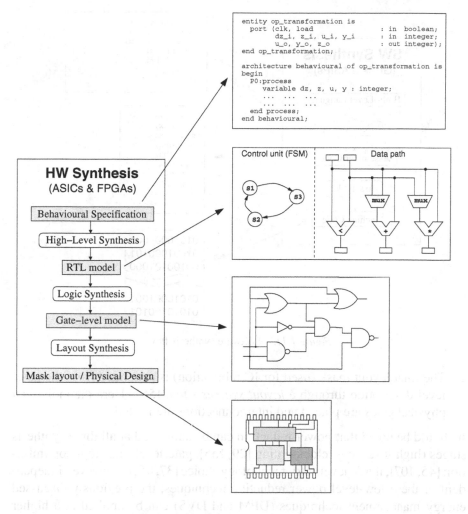

```
entity op_transformation is
    port (clk, load                    : in  boolean;
            dz_i, z_i, u_i, y_i          : in  integer;
            u_o, y_o, z_o                : out integer);
end op_transformation;

architecture behavioural of op_transformation is
begin
    P0:process
        variable dz, z, u, y : integer;
        ...  ...  ...
        ...  ...  ...
    end process;
end behavioural;
```

Figure 1.10. Hardware synthesis flow

154, 155]. Figure 1.10 illustrates a possible hardware synthesis process that consists of three subsequent design steps.

(a) A *high-level synthesis tool* (or behavioural synthesis tool) [8, 134, 154] trans-
    forms a behavioural specification into a structural description at the register-
    transfer level (RTL). Here the individual components are represented by data
    paths which execute arithmetic operations under control of a control unit.

(b) The RTL description (e.g. in structural VHDL) is then translated into a gate-
    level representation using a *logic synthesis tool* [9, 10]. In this stage of the
    design, the control unit as well as the data path are structurally represented
    as netlists of logic gates.

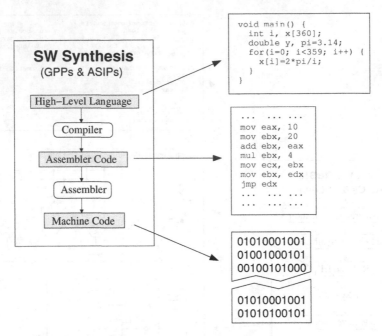

Figure 1.11.   Software synthesis flow

(c) The final layout mask (used for IC fabrication) is generated from the gate-level description through a *layout synthesis tool* [56]. Here the individual physical gates are placed and interconnections are routed.

It should be noted that power reduction can be addressed at all three synthesis stages (high-level: e.g. clock-gating [29, 155], gate-level: e.g. logic optimisation [45, 107], mask-level: e.g. technology choice [37, 45]). However, independent of these low-level power reduction techniques, the previously discussed energy management techniques (DPM and DVS) can be applied at a higher level of abstraction (system-level) to further improve the savings in energy. In general, the higher the level of abstraction at which the energy minimisation is addressed, the higher are the achievable energy saving [123].

Software Synthesis: Similarly to the hardware synthesis, all tasks that have been mapped to software programmable components have to be transformed from a high-level description (e.g. C/C++, JAVA, SystemC) into low-level machine code. A software translation hierarchy is shown in Figure 1.11 and consists of two steps:

(a) The initial specification in a high-level language is *compiled* into assembly code. This is carried out either using standard compilers, such as GCC [1, 2], or using specialised compilers that are optimised towards specific processor types (e.g. DSPs) [93]. The goal of the optimisation is the effective assignment of variables to registers such that operations can be performed without "time consuming" memory accesses.

(b) Once an optimised assembly code has been generated, the low-level code generation is carried out by processor specific assemblers that translate the assembler code into executable machine code.

There exist also techniques for compiler-based power minimisation such as instruction reordering and reduction of memory accesses [94, 145, 146]. Further, sizeable power saving can be obtained through a careful algorithmic design at the source code level [139]. Clearly, such software power minimisation approaches and system-level energy management techniques do not exclude each other. In fact, for a most energy-efficient system design both techniques should be considered.

1.5 Book Overview

This work presents novel techniques and algorithms for the automated design of energy-efficient distributed embedded system. In particular, the energy reduction capabilities of dynamic voltage scaling (DVS) are investigated and analysed in the context of highly programmable embedded systems with strict performance and cost requirements. The remainder of this book is organised as follows. Chapter 2 provides a survey of the most relevant and related works and outlines the necessary background information that is helpful for the understanding of the discussed subject.

Chapter 3 introduces a technique for dynamic voltage scaling in distributed architectures that effectively reduces the energy dissipation of the embedded system. This technique addresses the energy management problem discussed in Section 1.3. The proposed approach considers the power variations inherent to the execution of different tasks, in order to increase the efficiency with which DVS can be applied.

Based on this DVS technique, Chapter 4 introduces a new co-synthesis approach for distributed embedded system that potentially contain voltage-scalable components. Application mapping and activity scheduling are optimised towards the effective utilisation of DVS, i.e., towards energy reduction. This optimisation simultaneously aims at the identification of solution candidates that fulfil the imposed timing constraint and reduced the system cost.

Chapter 5 further extends the proposed co-synthesis approach towards the design of multi-mode embedded systems which integrate several different applications into a single device. The introduced multi-mode co-synthesis aims at energy-efficiency as well as cost effective utilisation of the hardware components. It is demonstrated that substantial energy savings can be achieved without modification of the underlying hardware architecture, even when neglecting DVS.

Many real-world applications exhibit control-intensive behaviour on top of the transformational data flow. Such systems can be modelled through conditional task graphs. A dynamic voltage scaling and scheduling technique for such application types is introduced in Chapter 6.

The techniques introduced in the preceding chapters and their algorithmic implementations have been combined into a new prototype co-synthesis tool for energy-efficient embedded systems. This tool is introduced in Chapter 7 and its usage is demonstrated using a real-life smart-phone that merges a cellular GSM phone, a digital camera, and an MP3-player into one device. Chapter 8 concludes the presented work and outlines potential areas of future research.

Chapter 2

BACKGROUND

Reducing power consumption has emerged as a primary design goal, in particular for battery-powered embedded systems. Low power design techniques for digital components have been intensively investigated over the last decade [28, 108, 116, 122, 147, 155]. These techniques focus mainly on the optimisation of a single hardware component in isolation. However, embedded systems are often far more complex than single components — they consist of several interacting heterogeneous components. Here the interrelation between the different processors and hardware blocks should be carefully considered during the synthesis in order to achieve an energy-efficient design. Two techniques that can be used for energy minimisation of distributed embedded systems are: dynamic power management (DPM) [23] and dynamic voltage scaling (DVS) [80, 152]. These system-level energy management techniques achieve energy reductions by selectively switching off unused components (DPM) or by scaling down the performance of individual components in accordance to temporal performance requirements of the application (DVS).

The aim of this chapter is to introduce the sources of power dissipation within distributed embedded systems and to outline how energy management techniques can be applied to reduce the dissipated energy (Section 2.1–2.3). Furthermore, an overview of the most relevant previous work is given, differentiating between general co-synthesis approaches without energy minimisation and co-synthesis approaches with energy minimisation (Section 2.4).

2.1   Energy Dissipation of Processing Elements

The power P_{PE} dissipated by computational components (CPUs, ASIPs, FPGAs, ASICs) of an embedded system, i.e. processing elements, is caused by two distinctive effects. First, *static* currents that occur whenever the processing

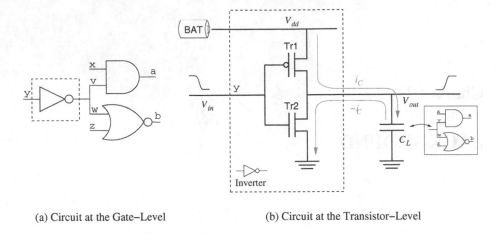

(a) Circuit at the Gate–Level   (b) Circuit at the Transistor–Level

Figure 2.1. Dynamic power dissipation of an inverter circuit [37]

element is switched on, even when no computations are carried out on this unit. Second, active computations cause switching activity within the circuitry that results in *dynamic* power dissipation whenever computations are performed. Accordingly to both sources the total power dissipation of processing elements is given by:

$$P_{PE} = P_{static} + P_{dynamic} \tag{2.1}$$

Both static and dynamic power dissipations can be further subdivided into two power components, each [37]:

$$P_{PE} = \underbrace{P_{leak} + P_{bias}}_{P_{static}} + \underbrace{P_{sc} + P_{sw}}_{P_{dynamic}} \tag{2.2}$$

The static power P_{static} has two parts: leakage power P_{leak} and bias power P_{bias}. While the dynamic power $P_{dynamic}$ consists of short-circuit power P_{sc} and switching power P_{sw}. Out of these four source of power dissipation, switching power P_{sw} is currently the dominant one which accounts for approximately 90% of the total PE power consumption [38]. Accordingly, the following discussion concentrates on this portion of the total power dissipation, simply referred to as power or dynamic power. It should be noted, however, that with shrinking feature size ($< 0.07 \mu m$) and reduced threshold voltage levels, the leakage currents become additionally an important issue [31, 37, 130].

Switching power P_{sw} is dissipated due to the charging and discharging of the effective circuit load capacitance C_L (parasitic capacitors of the circuit gates). To clarify the source of switching power P_{sw}, consider the simple gate-level circuit shown in Figure 2.1(a) and in particular the inverter gate shown in Figure 2.1(b). This inverter undergoes the following transitions. First, the input signal y is set to high (1), i.e., Tr1 is open (not conducting) while Tr2 is conducting. Accordingly, the circuit load capacitance C_L is discharged since Tr2 pulls the capacitance to ground. The load capacitance C_L represents the intrinsic capacitance of the inputs v and w of the AND and NOR gates. Now consider a transition from high (1) to low (0) at the input of the inverter y. In this case the transistor Tr2 is open and Tr1 connects the capacitance C_L to the supply voltage V_{dd} source, charging C_L via i_C. The power dissipated by this transition is given by:

$$P_{sw} = V_{dd} \cdot i_C \qquad (2.3)$$

where the dynamic current i_C changes according the dynamic voltage on the output:

$$i_C = C_L \cdot \frac{\partial V_{out}}{\partial t} \qquad (2.4)$$

Therefore, energy is transferred from the power supply to the load capacitance C_L. However, it can be observed that a transition from low to high at the input does not draw any current from the source, but instead discharges the load capacitance C_L via Tr2. This indicates that power, from the battery point of view, is only dissipated during output transitions from 0 to 1, i.e., when the load capacitance is charged. According to the above given observation, the energy consumption of the circuit is solely caused by transitions from low to high at the output of the gate. The dissipated switching energy [1] of one clock cycle, which takes a time of T, can be calculated as [37]:

$$E_{sw}^{0 \to 1} = \int_0^T P_{sw}\, dt = V_{dd} \cdot \int_0^T i_C\, dt = C_L \cdot V_{dd}^2 \qquad (2.5)$$

where the time $T = 1/f$ (period of on clock cycle) depends on the operational frequency f at which the circuit is clocked.

Although the above considerations were restricted to a single inverter gate, the same observations hold for more complex circuits, such as microprocessors [36]. As a result, the total energy E_{dyn}^{τ} drawn from the batteries by a PE performing a computational task τ depends additionally on the number of clock cycles N_C needed to execute this task and the switching activity α. Therefore, the total energy E_{dyn}^{τ} is given by:

$$E_{sw}^{\tau} = N_C^{\tau} \cdot \alpha \cdot C_L \cdot V_{dd}^2 \qquad (2.6)$$

Dividing Equation (2.6) by the execution time of the task $(T \cdot N_C^{\tau})$, the well known equation for power dissipation due to switching can be derived [37]:

$$P_{sw}^{\tau} = E_{sw}^{\tau} \cdot \frac{1}{T \cdot N_C} = \alpha \cdot C_L \cdot V_{dd}^2 \cdot f \qquad (2.7)$$

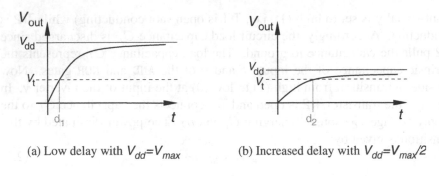

(a) Low delay with $V_{dd}=V_{max}$ (b) Increased delay with $V_{dd}=V_{max}/2$

Figure 2.2.   Supply voltage dependent circuit delay

Considering Equations (2.6) and (2.7) leads to very interesting conclusions. If we assume that the load capacitance C_L is a constant given by the complexity of the design and the circuit technology, and further the switching activity is a constant depending on the computational task then:

Although decreasing the operational frequency f leads to a reduction in power dissipation (Equation (2.7)), it does not reduce the energy dissipation (Equation (2.6)), which is important for battery-lifetime. A simple example can be used to illustrate this. Consider a computation that requires $10ms$ on a PE running at $10MHz$ and dissipating $200mW$. This computation requires an energy of $200mW \cdot 10ms = 2mJ$. If the operational frequency is reduced from $10MHz$ to $5MHz$, the power consumption of the processor decreases to $100mW$, according to Equation (2.7). Nevertheless, the frequency reduction increases the computational time from $10ms$ to $20ms$. Therefore, the dissipated energy remains unchanged $100mW \cdot 20ms = 2mJ$. Hence, the only possibility to reduce the energy consumption is to reduce the circuit supply voltage V_{dd}. Of course, reducing the supply voltage necessitates the reduction of the frequency in order to ensure correct operation, as it will be shown in the following.

When reducing the supply voltage of a digital circuit, the time required for gate signals to settle is prolonged, which, in turn, increases the circuit delay [37]. The source of this increased circuit delay is simplified shown in Figure 2.2. The two sub-figures show the gate output voltage V_{out} over time during a transition from low to high. While the Figure 2.2(a) corresponds to a transition at maximal supply voltage, Figure 2.2(b) shows a transition at reduced voltage. Subsequent gates recognise the signal as high as soon as the output voltage V_{out} exceeds the threshold voltage V_t. It can be seen that the reduction of the supply voltage V_{dd} leads to longer charging times until the threshold is reached (compare d_1 with d_2).

The circuit delay d, which is inverse proportional to the operational frequency f at which the component is clocked, can be approximated (error $< 10\%$) as [34, 37]:

$$d \propto \frac{1}{f} \approx k_d \cdot \frac{V_{dd}}{(V_{dd} - V_t)^2} \qquad (2.8)$$

with the technology dependent constant k_d which is given by:

$$k_d = \frac{C_L}{2 \cdot W \cdot v_{sat} \cdot C_{OX}} \qquad (2.9)$$

where the constants W, v_{sat}, and C_{OX} denote width of the sub-micron CMOS device, velocity saturation, and gate capacitance, respectively, which limit the charging current of load capacitance C_L. Based on the energy equation (2.6) and the circuit delay equation (2.8), the normalised energy/delay trade-off can be derived as:

$$E^*(d^*) = \frac{E_{max}}{V_{max}^2} \cdot \left(V_t + \frac{V_0}{2d^*} + \sqrt{\left(V_t + \frac{V_0}{2d^*} \right)^2 - V_t^2} \right)^2 \qquad (2.10)$$

where E_{max} and V_{max} denote the nominal values of energy dissipation and supply voltage, respectively. The normalised delay $d^* = d(V_{dd})/d(V_{max})$ is represented by the circuit delay at the scaled voltage $d(V_{dd})$ over the minimal circuit delay at nominal voltage $d(V_{max})$. While the constant V_0 is given by:

$$V_0 = \frac{(V_{max} - V_t)^2}{V_{max}} \qquad (2.11)$$

Please note that Equation (2.10) differs from the energy/performance trade-off equation given in [71]. The reason for this can be found in the fact that the considered PEs employ lower level power reduction techniques, e.g., gated clocks that avoid switching in unused circuit parts. Based on Equation (2.10), the energy-delay trade-off curve given in Figure 2.3 shows the normalised energy dissipation dependent on the normalised circuit delay for two cases: (a) keeping the supply voltage fixed and (b) dynamically adjusting the supply voltage. For instance, executing a task at $10MHz$ instead of $30MHz$ reduces the energy dissipation to approximately 33% of the nominal value when the supply voltage is lowered accordingly. On the other hand, if the supply voltage is kept fixed, the energy dissipation remains constant at the nominal value (see Equation (2.6)).

Although the above discussion is mainly concerned with the reduction of the *dynamic* power consumption, reducing the supply voltage V_{dd} also reduces the leakage power consumption, which is given by [102]:

$$P_{leak} = V_{dd} \cdot K_1 \cdot e^{K_2 \cdot V_{dd}} \cdot e^{K_3 \cdot V_{bs}} + |V_{bs}| \cdot I_{Ju} \qquad (2.12)$$

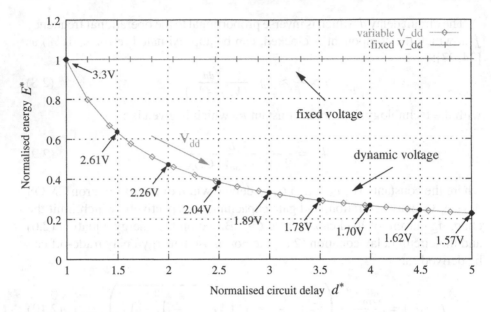

Figure 2.3. Energy versus delay function using fixed and dynamic supply voltages (considering $V_{max} = 3.3V$ and $V_t = 0.8V$)

where V_{bs} is the body bias voltage and I_{Ju} represents the body junction leakage current. The parameters K_1, K_2 and K_3 denote circuit technology dependent fitting constants. As we can observe, the leakage power P_{leak} depends linearly as well as exponentially on the supply voltage V_{dd}. Hence, reducing the supply voltage decreases the power consumption. Nevertheless, since constant K_2 is in general smaller than K_3, reducing the body bias voltage V_{bs} is more effective in limiting the leakage currents. Please note that the threshold voltage V_t depends on the body bias voltage, given as:

$$V_t = V_{th1} - K_4 \cdot V_{dd} - K_5 \cdot V_{bs} \tag{2.13}$$

where V_{th1}, K_4, and K_5 are constants for a given technology. Hence, scaling V_{bs} to reduce leakage will increase the circuit delay d, which is equivalent to a reduced performance as given in Equation (2.8). Interesting works on leakage power reduction can be found, for example, in [17, 86, 102].

2.2   Energy Minimisation Techniques

Having introduced the fundamentals of power dissipation in digital circuity, this section outlines two energy reduction techniques that have received considerable attention from academia and industry: dynamic voltage scaling and dynamic power management.

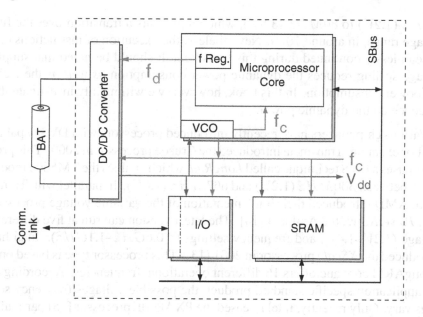

Figure 2.4. Block diagram of DVS-enabled processor [36]

2.2.1 Dynamic Voltage Scaling

A relatively new energy minimisation technique, which leverages the energy-delay trade-off described above (Section 2.1), is dynamic voltage scaling (DVS) [80, 152, 161]. DVS-enabled processors have the ability to dynamically change their supply voltage and operational frequency settings during run-time of the application. Hence, temporal performance requirements of applications can exploit the energy/delay trade-off to reduce energy dissipation. Figure 2.4 shows a block diagram of a typical DVS-enabled processor [36]. A microprocessor core carries out the required computations. This processing unit is connected through a system bus to the static memory (cache unit) and the I/O bus interface. The heart of this system, which enables a dynamic voltage selection, consists of a DC/DC voltage converter, a specialised frequency register, and a voltage controlled oscillator (VCO). Supply voltage and operational frequency are changed by writing the desired frequency f_d into the frequency register, i.e., these changes are carried out under software control. Upon writing the desired frequency into the register, the DC/DC converter compares this frequency with the current frequency f_c (which clocks the microprocessor core, cache, and I/O interface) and either increases or decreases the supply voltage V_{dd}. According to the changed voltage, the VCO adapts the system clock to a higher or lower frequency f_c. Certainly, the whole voltage scaling process requires a finite time. Typical transition times are in the range of tenths of microseconds. For instance, the prototype processor introduced in [36] is able to switch from

$5MHz$ (1.2V) to $80MHz$ (3.8V), which represents a transition over the full voltage range, in around $70\mu s$. Nevertheless, the execution of instructions can be seamlessly continued during this transition. It should be noted that supply voltage scaling reduces the dynamic power consumption *as well as* the leakage power consumption. In this book, however, we will mainly investigate the influence on the dynamic power.

Various chip makers have recently introduced processors with DVS capability. For example, Transmeta introduced the Crusoe processor in 2000. This processor uses a DVS technique called *LongRun*, which enables the TM5600 model to run between $300MHz$ (1.2V) and $667MHz$ (1.6V). In parallel with Transmeta, AMD introduced their implementation of the variable voltage processor with *PowerNow!*, the Athlon 4 [14]. The latest version can run at five different voltage (1.2V–1.4V) and frequency settings ($\leq 0.6GHz$–$1.1GHz$). Intel has introduced the *XScale* processor in 2001 [13]. This processor type is based on a StrongARM core and offers 16 different operational frequencies. According to the application specific standard product, the possible voltages/frequency settings vary. Only recently, Intel released the PXA800F processor [15] particular suitable for cellular phones. Based on an ARM core with *XScale*, this processor has the ability run at $104MHz$ and $312MHz$. The increasing availability of DVS processors adds credibility to the practicability of the DVS technique. Certainly, DVS is becoming an important energy reduction technique.

2.2.2 Dynamic Power Management

Unlike DVS, dynamic power management (DPM) is already around for quite a while. The main strategy of DPM is the shutdown of idle system components [26, 97]. An advantage of DPM is its generality [23], which allows its usage not only for digital circuitry, but also for other system components such as displays, hard drives, and analogue circuits. DPM approaches often differ in the employed shutdown policies (or power management policies). For example, in its most aggressive strategy components are switched off immediately when they become idle. A second strategy is the timeout-based policy, which switches off components after a fixed idle interval. This policy is well known from the advanced power management (APM) widely used in today's notebook computers. Nevertheless, since the restart of a component involves a time and power overhead to restore its fully functional state, such greedy policies might not result in power savings or may even increase the dissipated power [24]. Therefore, a careful consideration of the applied policies is necessary to achieve the highest possible power savings [23–25, 97]. The problem with most real-life systems is the uncertainty with which future events occur. The quality of a power management policy depends therefore on the accuracy with which the future behaviour of the system can be predicted, in order to start-up currently inactive components or to shut down currently active components at the right moment.

2.2.3 Dynamic Power Management versus Dynamic Voltage Scaling

DVS and DPM are both useful techniques that help to reduce the energy dissipation of an embedded system. However, one valid question, which has not been discussed so far, is: "Why should one use a rather complex technique like DVS to slow down processing elements if it is possible to switch off components during idle intervals?" This question can be illustratively answered using a simple example. For clarity reasons, the timing overheads for voltage scaling and power management are neglected here. Consider the following situation. A processing element (processor) performs a certain task τ_x in $20ms$ and dissipates a power of $500mW$ when running with $33MHz$ at a nominal supply voltage of $3.3V$ and a threshold voltage of $0.8V$. To meet the performance requirements of the application, the task needs to be repeated every $30ms$. Thus, between each consecutive execution of the task there are $10ms$ of idleness during which the processor can be deactivated (DPM). Under these circumstances the execution of task τ_x results in an energy consumption of $500mW \cdot 20ms = 10mJ$. Figure 2.5 illustrates this situation.

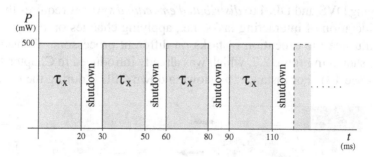

Figure 2.5. Shutdown during idle times (DPM)

Now consider the execution using voltage scaling. The repetition time of $30ms$ allow to prolong the execution time of task τ_x from $20ms$ to $30ms$. That is, instead of shutting down the processor during the $10ms$ of idleness, the processor's performance is reduced from $33MHz$ to $22MHz$, since $30ms/20ms = 1.5$ and $33MHz/1.5 = 22MHz$. The lower frequency allows the reduction of the supply voltage from $3.3V$ to $2.61V$, according to Figure 2.3 and the tolerable increase in circuit delay of $33MHz/22MHz = 1.5$. Based on Equation (2.8), the exact voltage can be obtained by considering the ratio between the delays at $33MHz$ and $22MHz$. At this voltage and frequency values the processing element dissipates a power of $209.8mW$ (see Figure 2.6). Similar to the voltage calculation, the exact value can be obtained from Equation (2.7), using the power ratio between the operation at $33MHz$ and $22MHz$. Thus, the energy

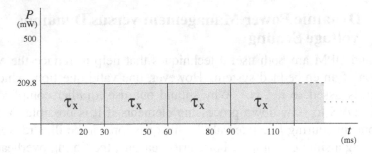

Figure 2.6.   Voltage scaling to exploit the slack time (DVS)

consumption is given by $209.8mW \cdot 30ms = 6.29mJ$, a reduction of 32.1% compared to the $10mJ$ dissipated when using DPM. This simple example has shown that the energy efficiency of DVS is superior when compared to DPM. In fact, DVS always performs better than DPM, whenever both techniques are applicable [76].

2.2.4   DVS and DPM in Distributed Embedded Systems

Applying DVS and DPM to *distributed embedded systems* requires the careful consideration of interacting tasks, i.e., applying changes on one processor might influence the execution of tasks on different processors. Consider the schedule shown in Figure 2.7, which was already introduced in Chapter 1 (Figure 1.9, Page 14). For instance, by slowing down CPU 2 during the execution

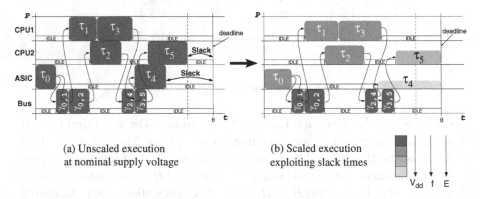

(a) Unscaled execution   (b) Scaled execution
at nominal supply voltage   exploiting slack times

Figure 2.7.   Combination of dynamic voltage scaling and dynamic power management

of task τ_2, the communications $\gamma_{2,4}$ and $\gamma_{3,5}$ are delayed, due to dependencies. This delay further influences the earliest possible start times of the tasks τ_4 and τ_5. These interdependencies further complicate the voltage scaling processes

[20, 68, 99, 133], compared to single processor systems. Another observation which can be made from Figure 2.7(b) is that even after applying DVS to exploit the available slack there remain some idle periods in the system schedule. Clearly, further lowering the voltage and frequency would result in missed deadlines θ, i.e., infeasible solutions. However, it is possible to switch off idle components during idle times, without influencing the schedule. In order to achieve a high degree of reduction, dynamic voltage scaling and dynamic power management should be considered together. However, due to the higher efficiency of DVS, DPM should be applied after DVS.

For DVS and DPM to be useful it is obligatory that the system experiences idle times (times where a certain components do not carry out an useful task) and slack times (times where a reduced system performance can be tolerated). Such idle and slack times can be found in distributed embedded systems due to four main reasons:

(a) It is often the case for a given application to show various degrees of parallelism, i.e., not all PEs will be utilised constantly during run-time. See, for instance, the schedule in Figure 2.7(a).

(b) The performance of the allocated architecture cannot be adapted perfectly to the application needs, since the allocation of "performance" is not given as a continuous range, but is rather quantised.

(c) The allocated architecture is commonly over-designed to allow an incremental design process, i.e., designers try to decrease development time through the reuse of the hardware architectures over several product generations. This is only possible by leaving enough "performance headroom" for future applications [120].

(d) Schedules for hard real-time systems are constructed by considering worst-case execution times (WCETs), however, actual execution times of tasks during operation are, for most of their activations, smaller than their WCETs [137].

2.3 Energy Dissipation of Communication Links

The preceding sections mainly concentrated on the power consumption of processing elements and suitable energy management techniques. However, in distributed heterogeneous system with interacting components, the transfer of data between processing elements additionally contributes to the overall energy consumption. Fast communication is essential to avoid undesired contention of processing elements. Therefore, wide system buses (8 to 128 bit) and high-speed serial buses (CAN bus, I^2C bus, USB, Gigabit Ethernet, Firewire, etc.) have become commonplace. With each transfer of data over the communication links (CLs), the line capacitance is charged and discharged, drawing a current from the I/O pins of the processing elements. The power dissipated by these currents is given by:

$$P_{CL} = \beta \cdot C_{bus} \cdot f_{bus} \cdot V_{tr}^2 \qquad (2.14)$$

where $\beta \cdot C_{bus}$ is the effectively switched load capacitance of the bus lines, V_{tr}^2 is the operational voltage of the bus, and f_{bus} is the operational frequency at which bits are transfered over the bus. Important for the battery-lifetime, however, is the drawn energy:

$$E_{CL}^\gamma = N_{tr}^\gamma \cdot \beta \cdot C_{bus} \cdot V_{tr}^2 \qquad (2.15)$$

where N_{tr}^γ denotes the number of bus cycles needed by communication γ. One possibility to reduce this energy dissipation is to encode the data before transferring it. Such bus-encoding techniques have been investigated with the aim of reducing the switching activity β [22, 114, 141]. Unlike the supply voltage of processing elements, the transmission voltage V_{tr}^2 cannot be reduced easily due to reliability issues, that is, environmental noise potentially corrupts data during transfers at low voltages. Nevertheless, many communication interfaces allow operation at different communication speeds (Ethernet at 10Mb/s, 100Mb/s, and 1Gb/s). In [96], Liu *et al.* present a technique that appropriately selects the speed of processors as well as communication links, in order to reduced the overall energy consumption. Furthermore, DPM can be equally applied to CLs as to PEs, i.e., during intervals where no data is transfered the CLs can be switched off.

2.4 Further Readings

This section briefly reviews the most relevant works in the area of co-synthesis for embedded systems, distinguishing between co-synthesis approaches that neglect issues related to power and more recent co-synthesis approaches that aim at the reduction of power consumption. Due to the topic of this book, special emphasis is placed upon the latter. Accordingly, this section gives the interested reader some starting points for further readings.

2.4.1 Co-Synthesis without Energy Minimisation

Hardware/software co-synthesis is a field of active research since the early 90's and several enlightening surveys have been published ever since [55, 104, 105, 142, 156]. Initially, most co-synthesis approaches targeted an architecture consisting of a single general-purpose processor connected to one or a few ASICs. Therefore, the main goal was to split the application between slow-but-cheap software and fast-but-expensive hardware, to achieve the highest possible performance without exceeding a given cost constraint [52, 53, 69, 89, 132, 153]. In these approaches special attention is given to different heuristic algorithms that solve the computational expensive partitioning problem, while issues related to power consumption are neglected. Ernst *et al.* [53] presented a simulated annealing based partitioning which starts from an all-in-software solution. Nodes are moved towards hardware until a given timing constraint is satisfied. Gupta and DeMicheli [69, 70] proposed a heuristic that iteratively moves nodes from hardware to software to reduce cost without the violation of imposed timing constraints. Kalavade and Lee [83] introduced a technique that utilises a global-criticality/local-phase measure to partition the system via a list scheduling approach [16]. Eles *et al.* [52] investigated a tabu-search partition-ing algorithm, and a comparison to simulated annealing is provided. A genetic algorithm for hardware/software partitioning was developed in [132]. A com-parison between simulated annealing, tabu-search, and genetic algorithm for one-CPU-one-ASIC architectures can be found in [153]. Other techniques that have been investigated include branch-and-bound algorithms [40] and dynamic programming [89].

More recent co-synthesis approaches target distributed heterogeneous archi-tectures, i.e., architectures which potentially contain several processing ele-ments and communication links of different types. The first work in this area was done by Prakash and Parker [121]. They formulated and solved the problem using mixed integer linear programming. In a similar vein, Bender [21] applied mixed integer linear programming. In [157], Wolf developed a greedy heuristic which initially allocates a processing element for each task. Lightly loaded PEs are removed iteratively by re-assigning task to other components. Teich *et al.* [143] applied an evolutionary approach to the system-level co-synthesis problem using a graph-based mapping model. Oh and Ha [112] introduced a co-synthesis framework based on heterogeneous multiprocessor scheduling [111]. Pop *et al.* [119, 120] addressed co-synthesis problems that are typical to time-triggered communication protocols and incremental design processes. Madsen and Bjørn-Jørgensen [101] investigated the co-design of embedded systems under memory constraints.

2.4.2 Co-Synthesis with Energy Minimisation

All the approaches introduced in the previous section greatly neglect the optimisation of energy consumption. However, the recent development of the portable-application market has intensified the interest in system-level design techniques for energy-efficient embedded systems. The first approach that targeted the reduction of power dissipation throughout the co-synthesis pro-cess was proposed by Dave *et al.* [41]. They developed a constructive al-gorithm based on energy levels, which makes the mapping of tasks energy sensitive. Dick and Jha [49] reported a multi-objective genetic algorithm based co-synthesis approach. This framework simultaneously optimises the embed-ded system for cost, power consumption, and timing behaviour. An excellent overview of scheduling and synthesis techniques for low power can be in [82].

Most recently, co-synthesis approaches started to integrate the consideration of energy management techniques. This allows to optimise the embedded system towards the effective utilisation of DPM and DVS, hence, the reduction of energy consumption. In [74], Henkel introduces a low-power hardware/software partitioning approach for core-based systems. To avoid unnecessary high switching activity, the application is carefully partitioned into cores that can be selectively switched off. This technique is particularly suitable for designs in which no clock gating is applied. Lu *et al.* [98] presented a greedy on-line scheduling technique that optimises the execution order of tasks towards the utilisation of DPM. The main idea is to cluster the execution of tasks instead of scattering them, so that high shutdown overheads in terms of power and time can be avoided. Similarly, Brown *et al.* [33] developed a buffer insertion and scheduling technique for distributed systems that allows to reduce the shutdown overheads to improve the utilisation of DPM. In [76], Hong *et al.* introduced a design methodology for low power core-based real-time systems-on-a-chip. A variable voltage scheduling technique for single processor systems was proposed. Luo and Jha [99] developed a combined scheduling technique for periodic executing tasks with dependencies as well as aperiodic tasks. Dynamic voltage scaling is considered in this scheduling context. Dependent tasks are scheduled statically. The schedulability of aperiodic tasks as well as the utilisation of DVS are improved by distributing available slack time evenly among the processing elements of a distributed architecture. The online scheduler serves the aperiodic tasks and considers resource reclaiming. The same authors further enhanced their approach towards a battery-aware scheduling with the aim to improve the battery discharge profile [100]. This technique is based on the fact that a flattening out the discharge curve it is possible to prolong battery-lifetime [92, 115, 124]. Gruian and Kuchcinski [67, 68] extended a dynamic list scheduling heuristic to support DVS by making the priority function energy aware. In each scheduling step the energy-sensitive task priorities are re-calculated. If a scheduling attempt fails (exceeded hard deadline), the priority function is adjusted and the application is re-scheduled. Bambha *et al.* [20] presented a hybrid search strategy based on simulated heating, in order to derive an energy-efficient voltage selection for individual tasks. Mapping and scheduling are based on a dynamic list scheduling approach [138]. Recently, an integer programming (IP) was proposed by Zhang *et al.* , which can been shown to solve the voltage scaling problem optimally in pseudo-polynomial time [163], under the assumption that continuous voltage-scalable processors are given. The work introduced by Andrei *et al.* [17] proposes a non-linear programming solution for the continuous voltage scaling problem under consideration of scaling overheads in terms of energy and time. Furthermore, the discrete voltage scaling problem is demonstrated to be NP-hard (with and without the consideration of scaling overheads), and the an integer programming solution is introduced to solve the problem including the scaling overheads.

Many DVS approaches for heterogeneous distributed embedded systems have assumed a *fixed power dissipation* of processing elements during the energy minimisation. In reality this might not be true. For instance, low-level power minimisation techniques, such as clock gating [28, 29], result in power variations during the execution of an application. The problem of considering these power variations during dynamic voltage scaling is addressed in Chapter 3.

Other noticeable approaches, which target energy reduction, are the works by Chung *et al.* [39] and Yang *et al.* [160]. Chung *et al.* [39] achieved energy efficiency by leveraging information regarding the execution time variations, which is supplied to mobile terminals by the contents provider. That is, the contents provider estimates the performance required by the mobile terminals. This estimation is transmitted to the mobile terminals, which accordingly adapted there performance to save energy. The approach presented by Yang *et al.* [160] uses a two-phase scheduling method. In the first stage, which is performed off-line (during design time), a Pareto-optimal set of schedules is generated. These schedules provide different execution time/energy trade-offs. During run-time, a run-time scheduler selects points along the Pareto set, in order to account for the dynamic behaviour of the application.

2.5   Concluding Remarks

This chapter has introduced the sources of power dissipation within distributed embedded systems. Two energy reduction techniques, namely dynamic power management and dynamic voltage scaling, have been outlined, and the advantage of DVS over DPM has been highlighted. Furthermore, an overview of most relevant previous works in the area of co-synthesis has been given.

Notes

1 In order to prevent a confound usage of the terms *power* and *energy*, the following definition is used throughout this book. The term *power dissipation* refers to the physical value power, while the terms *power consumption*, *energy dissipation*, and *energy consumption* refer to physical value energy.

Chapter 3

POWER VARIATION-DRIVEN DYNAMIC
VOLTAGE SCALING

Dynamic voltage scaling is a powerful technique to reduce the energy consumption of processing elements within embedded systems. By simultaneously scaling supply voltage and operational frequency it becomes possible to trade off between performance and energy dissipation. This effect can be used to exploit the temporal performance requirement of an application, in order to save energy. Many approaches to DVS in distributed systems assume that the power dissipation of processing elements is independent of the executed instructions [20, 68, 99]. Therefore, the voltage selection is carried out considering a constant power dissipation, in the following referred to as *fixed power model*. In practice this may not be true. For instance, modern IC designs often make use of low-level power minimisation techniques, such as clock gating to stop the switching activity in un-utilised blocks of the circuit [45, 146]; gated clocks are also used for DVS-PEs [34]. Under such circumstances the power dissipation can vary considerably during the execution of different functions [32]. This chapter presents a voltage scaling technique that takes this power variation effect into account, in order to increase the potential energy savings. The voltage scaling strategy is based on an energy-gradient driven heuristic, and the concept of mapped-and-scheduled task graphs is used to account for task and communication dependencies in a fast and effective way.

The remainder of this chapter is organised as follows. Section 3.1 motivates the need for a refined voltage selection that accounts for power variations. Section 3.2 introduces a new voltage scaling technique which accounts for power variations. Experimental results are presented in Section 3.3. Finally, concluding remarks are given in Section 3.4.

35

3.1 Motivation

This section motivates the consideration of power variations during the voltage scaling. The aim of this consideration is to improve the efficiency with which DVS can be applied.

Variations in the power dissipated by the individual tasks of an embedded system are caused due to the following two reasons: Firstly, modern IC designs make heavy use of gated clocks, i.e., unused circuity is selectively "turned off" by stopping the clock signal to it [45, 146]; this holds also for DVS-PEs [34]. Thereby, unnecessary charging and discharging of the circuit load capacitance is avoided, which, in turn, results in lower power dissipation (Equation (2.7), page 21). For example, in the case of a general-purpose processor (GPP), including an integer unit and a floating point unit (FPU), it is not desirable to keep the FPU active if only integer instructions are executed. Hence, the clock signal to the FPU can be gated (stopped) during the execution of integer instructions, which will nullify the switching activity in the FPU. Thus, different tasks (different use of instructions) dissipate different amounts of power on the same processing element. In the case of an ARM7TDMI processor the supply current varies between $5.7mA$ and $18.3mA$, depending on the functionality which is carried out [32]. The second reason for the power variation effect is typical for core-based designs, where several different cores reside together on a single chip. Consider an ASIC accommodating four different cores: a FIR filter, an IDCT algorithm, a DES encrypt/decrypt unit, CORDIC (coordinate rotation digital computing) algorithm. Clearly, these four cores vary considerably in complexity. For instance, logic synthesis results show that the FIR filter requires approximately three times less area than the DES crypto unit [11]. This heterogeneity results certainly in different power dissipations, accordingly to which cores are active at a given time.

Taking this power variations into account during the voltage scaling improves the overall energy efficiency, since the available slack time is distributed more fairly among the tasks. To illustrate the influence of power variation effects on the voltage selection, a motivational example is given next. However, before starting with the example, it is necessary to define the term *energy-gradient*, which will be used throughout this chapter.

DEFINITION 3.1 *An energy-gradient ΔE_τ is defined as the difference between the energy dissipation of task τ with the execution time t_{exe} and the reduced energy dissipation (due to voltage and clock scaling) of the same task when extended by a time quantum Δt.*

Mathematically:

$$\Delta E_\tau = E_\tau(t_{exe}) - E_\tau(t_{exe} + \Delta t) \qquad (3.1)$$

where $E_\tau(t_{exe})$ and $E_\tau(t_{exe} + \Delta t)$ are calculated based on Equation (2.10). □

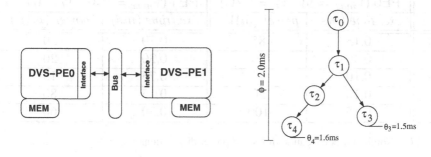

(a) Block diagram of the dynamic voltage scalable target architecture

(b) Task graph specification including period and deadlines

Figure 3.1. Architecture and specification for the motivational example

Motivational Example: Considering Power Variations during Voltage Scaling

The intention with this illustrative example is to motivate the consideration of power variation effects during the voltage scaling of heterogeneous distributed systems. This is done by using two different power models during the DVS optimisation: (a) the traditional *fixed power model* (assuming constant power) [20, 68, 99] and (b) a *power variation model* which accounts for the power dissipation of each task (as proposed in this chapter). Additionally, in order to address the voltage scaling problem in the presents of power variations a new energy-gradient voltage scaling strategy is introduced.

The starting point for applying any DVS technique is a scheduled (at nominal voltage and frequency) system specification consisting of tasks and communication, which are mapped onto an allocated architecture. In this simple example, the considered architecture is composed of two heterogeneous DVS-PEs (e.g. a Transmeta Crusoe [87] and a StrongARM with Xscale technology [13]), as shown in Figure 3.1(a). Each PE has its own local memory to store the different tasks that are mapped onto it. These processing elements are connected through a single bus. The system specification is given by the task graph shown in Figure 3.1(b), including repetition rate ($\phi = 2ms$) and deadline constraints ($\theta_4 = 1.6ms$ and $\theta_3 = 1.5ms$). Nominal supply voltage V_{max} and threshold voltage V_t for the two PEs are given in Table 3.1. This table further shows the nominal execution times and dynamic power dissipations of each task, according to their mapping (execution on PE0 or PE1). Furthermore, the transfer times and power dissipation of the communication activities are shown in Table 3.2, reflecting the inter-PE communications through the bus. Communications between tasks on the same PE (intra communications) are assumed to

task	PE0 ($V_{max} = 5V$, $V_t = 1.2V$)		PE1 ($V_{max} = 3.3V$, $V_t = 0.8V$)	
	exe. time (ms)	power (mW)	exe. time (ms)	power (mW)
τ_0	0.15	85	0.70	30
τ_1	0.40	90	0.30	20
τ_2	0.10	75	0.75	15
τ_3	0.10	50	0.15	80
τ_4	0.15	100	0.20	60

Table 3.1. Nominal task execution times and power dissipations

be instantaneous, and their power dissipation is neglected, as in most previous co-synthesis approaches [30, 41, 49, 89, 99, 118].

comm.	comm. time (ms)	power dis. (mW)
$\gamma_{0\rightarrow1}$	0.05	5
$\gamma_{1\rightarrow2}$	0.05	5
$\gamma_{1\rightarrow3}$	0.15	5
$\gamma_{2\rightarrow4}$	0.10	5

Table 3.2. Communication times and power dissipations of communication activities mapped to the bus

One possible mapping and scheduling of the system tasks and communications onto the underlying architecture is shown in Figure 3.2, which describes the power dissipation over time, that is, the power profile of PEs and CLs. It can be observed that PE0 accommodates task τ_0 and τ_4, while the remaining tasks (τ_1, τ_2, τ_3) are mapped to PE1. In accordance to this mapping, two inter-PE communications over the bus are required between tasks τ_0 and τ_1 as well as between tasks τ_2 and τ_4. The communication link CL0, connecting both PEs, shows these two communications, $\gamma_{0\rightarrow1} = (\tau_0, \tau_1)$ and $\gamma_{2\rightarrow4} = (\tau_2, \tau_4)$.

The dynamic system energy dissipation of this configuration at nominal supply voltage can be calculated using the dynamic power values and execution times given in Tables 3.1 and 3.2. The tasks mapped to PE0 (τ_0 and τ_4) consume an energy of $0.15ms \cdot 85mW + 0.15ms \cdot 100mW = 27.75\mu J$, while the tasks assigned to PE1 (τ_1, τ_2, τ_3) dissipate $0.3ms \cdot 20mW + 0.75ms \cdot 15mW + 0.15ms \cdot 80mW = 29.25\mu J$. Taking into account the communications on CL0 ($\gamma_{0\rightarrow1}$ and $\gamma_{2\rightarrow4}$), which consume $0.05ms \cdot 5mW + 0.10ms \cdot 5mW = 0.75\mu J$, the overall energy dissipation results in $27.75\mu J + 29.25\mu J + 0.75\mu J = 57.75\mu J$. Obviously, since the execution of task τ_3 finishes at $1.4ms$ and the task deadline is at $1.5ms$, a slack time of $0.1ms$ is available, as indicated in Figure 3.2. The same holds for task τ_4, which finishes its execution after $1.5ms$, leaving a slack

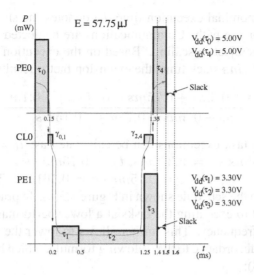

Figure 3.2. Power profile of a possible mapping and schedule at nominal supply voltage (no DVS is applied)

of $0.1ms$ until its deadline of $1.6ms$ is reached. These slacks can be used to extend the task execution times. Thus, the DVS-PEs can be slowed down by scaling the supply voltage and accordingly the clock frequency, following the relation given in Equation (2.8). Let us consider two cases for the identification of scaling voltages: (a) When a fixed power model is used (power variations are neglected, as in previous work [20, 68, 99]), i.e., all tasks mapped to the same PE are assumed to consume the same constant amount of power, and (b) a more generalised and realistic power model allowing for power variations among the tasks (as proposed in this work).

One approach to optimise the energy dissipation, which neglects the power profile, is to distribute the slack time evenly among the tasks, that is, in the given example each task is "stretched" using the same extension factor [68]. This is illustrated in Figure 3.3(a), where each task execution is extended using a factor of $e = 1.074$. For example, the execution of task τ_0 was extended to $0.161ms$ ($0.15ms \cdot 1.074$). Similarly, the remaining tasks. The extension factor can be calculated considering the longest path to the task with the smallest slack time. In the example at hand, both deadline tasks (τ_3 and τ_4) have a slack of $0.1ms$; hence, we have the choice (both possibilities would result in the same extension factor). Consider the path indicated in Figure 3.2, which involves the tasks τ_0, τ_1, τ_2, and τ_4. The extension factor e is given by:

$$e = \left(\left(\sum_{\tau \in \mathcal{T}_p} t_{nom}(\tau) \right) + t_S \right) / \sum_{\tau \in \mathcal{T}_p} t_{nom}(\tau)$$

where t_{nom} is the nominal execution time, t_S denotes the slack time, and $\mathcal{T}_p$ refers to all tasks on the path. Communications are neglected in this equation since they are not subject to scaling. Based on the execution values given in Table 3.1 and the $0.1ms$ slack time, the extension factor results in:

$$e = \frac{(0.15ms + 0.3ms + 0.75ms + 0.15ms) + 0.1ms}{0.15ms + 0.3ms + 0.75ms + 0.15ms} \approx 1.074$$

Thus, the extended task executions can be calculated as: $t_0 = 0.15ms \cdot e = 0.161ms$, $t_1 = 0.3ms \cdot e = 0.322ms$, $t_2 = 0.75ms \cdot e = 0.806ms$, $t_3 = 0.15ms \cdot e = 0.161ms$, and $t_4 = 0.15ms \cdot e = 0.161ms$. These execution times correspond to the schedule shown in Figure 3.3(a). In practice, extending a task is equivalent to executing the tasks at a lower performance, that is, at a lower operational frequency. This further allows to lower the supply voltages of PE0 and PE1 in accordance to the following formula, which has been derived from Equation (2.8):

$$V_{dd} = V_t + \frac{V_0}{2d^*} + \sqrt{\left(V_t + \frac{V_0}{2d^*}\right)^2 - V_t^2} \qquad (3.2)$$

where d^* denotes the normalised delay, which, in this example, is equal to the extension factor e. The constant V_0 is given by:

$$V_0 = \frac{(V_{max} - V_t)^2}{V_{max}} \qquad (3.3)$$

Thus, the reduced voltages of PE0 is calculated as:

$$V_{dd} = 1.2V + \frac{(5V - 1.2V)^2/5V}{2 \cdot 1.074} +$$
$$+ \sqrt{\left(1.2V + \frac{(5V - 1.2V)^2/5V}{2 \cdot 1.074}\right)^2 - (1.2V)^2}$$
$$V_{dd} = 4.788V$$

using the nominal supply voltage $V_{max} = 5V$ and the threshold voltage $V_t = 1.2V$ as given in Table 3.1. In the same way, the scaled supply voltage for PE1 can be calculated as $V_{dd} = 3.161V$, using $V_{max} = 3.3V$ and $V_t = 0.8V$. Adjusting the supply voltages of the PEs to theses levels, the task deadlines are still satisfied, and the power dissipations are reduced. According to Equation (2.7), the power dissipation of each task can be calculated using the following relation:

$$\frac{P_{V_{dd}}}{P_{V_{max}}} = \frac{\alpha \cdot C_L \cdot f_{V_{dd}} \cdot V_{dd}^2}{\alpha \cdot C_L \cdot f_{V_{max}} \cdot V_{max}^2} = \frac{1}{e} \cdot \frac{V_{dd}^2}{V_{max}^2} = \frac{1}{d^*} \cdot \frac{V_{dd}^2}{V_{max}^2} \qquad (3.4)$$

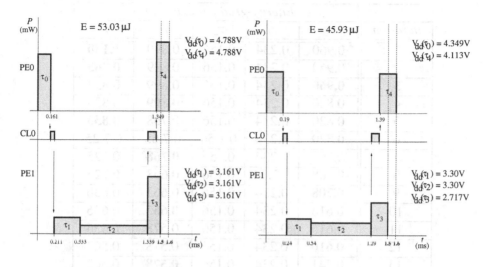

(a) Voltage scaled schedule using the fixed power model (traditional method)

(b) Voltage scaled using the power variation model

Figure 3.3. Two different voltage scaled schedules

considering that $\alpha \cdot C_L$ is constant for a given task and that relation between reduced operational frequency $f_{V_{dd}}$ and maximal operational frequency $f_{V_{max}}$ is equivalent to the inverse of the extension factor $1/e = f_{V_{dd}}/f_{V_{max}}$. Out of this, the individual power dissipations can be calculated as: $P_{Vdd}(\tau_0) = 85mW \cdot 1/1.074 \cdot (4.788V)^2/(5V)^2 = 72.57mW$, $P_{Vdd}(\tau_1) = 20mW \cdot 1/1.074 \cdot (3.161V)^2/(3.3V)^2 = 17.08mW$, $P_{Vdd}(\tau_2) = 12.81mW$, $P_{Vdd}(\tau_3) = 68.33mW$, and $P_{Vdd}(\tau_4) = 85.38mW$. Taking into account the extended execution times and the energy dissipated by communications, the total energy consumption of the schedule shown in Figure 3.3(a) results in: $E = (72.57mW \cdot 0.161ms + 17.08mW \cdot 0.322ms + 12.81mW \cdot 0.806ms + 68.33mW \cdot 0.161ms + 85.38mW \cdot 0.161ms) + (5mW \cdot 0.05ms + 5mW \cdot 0.10ms) = 53.03\mu J$. This is equivalent to a reduction of 8.2%.

Now consider the case when the generalised power model (allowing power variations) is employed during the identification of scaling voltages for the task executions. This optimisation is based on an energy-gradient that has been defined in Equation (3.1). For a simpler illustration of the method, a fixed time quanta size $\Delta t = 0.01ms$ is assumed, which is 10 times smaller than the available deadline slack ($0.1ms$). Having defined the time quantum size Δt, it is now possible to calculate an energy-gradient ΔE_τ for each task, using Equations (3.2) and (3.4)). For instance, consider task τ_0. The energy dissipation of this task at nominal supply voltage can be calculated as

iteration	Energy-gradient ΔE (μJ)				
	τ_0	τ_1	τ_2	τ_3	τ_4
1	0.960	0.234	0.156	0.899	**1.130**
2	0.960	0.234	0.156	0.899	**0.965**
3	**0.960**	0.234	0.156	0.899	0.833
4	0.820	0.234	0.156	**0.899**	0.833
5	0.820	0.234	0.156	0.768	**0.833**
6	**0.820**	0.234	0.1.56	0.7.68	0.7.25
7	0.708	0.234	0.156	**0.768**	0.725
8	0.708	0.2.34	0.156	0.663	**0.725**
9	**0.708**	0.234	0.156	0.663	0.636
10	0.616	0.234	0.156	**0.663**	0.636
11	0.616	0.234	0.156	0.578	**0.636**
12	**0.616**	0.234	0.156	0.578	0.562
13	0.541	0.234	0.156	**0.578**	0.562
14	0.541	0.234	0.156	0.507	**0.562**
15				**0.507**	
16				**0.451**	
extension	**0.04**	**0**	**0**	**0.06**	**0.06**

Table 3.3.   Evolution of the energy-gradients during voltage scaling

$E_0(0.15) = 0.15ms \cdot 85mW = 12.75\mu J$, using the values given in Table 3.1. Extending the execution of the task by $\Delta t = 0.01ms$ from $0.15ms$ to $0.16ms$ is equivalent to an extension factor of $0.16ms/0.15ms = 1.0667$. The scaled supply voltage of PE0 for this extension is $4.808V$ according to Equations (3.2). Running the task at this supply voltage would result in a power dissipation of $85mW \cdot 1/1.0667 \cdot (4.808V)^2/(5V)^2 = 73.69mW$ (using Equation (3.4)), and hence lead to an energy consumption of $73.69mW \cdot 0.16ms = 11.79\mu J$. Thus, the energy-gradient is given by $\Delta E_0 = 12.75\mu J - 11.79\mu J = 0.960\mu J$. The energy-gradients of the remaining tasks are calculated in the same way and result in: $\Delta E_1 = 0.234\mu J$, $\Delta E_2 = 0.156\mu J$, $\Delta E_3 = 0.899\mu J$, and $\Delta E_4 = 1.130\mu J$. Clearly, the task with the highest energy-gradient (in this case task τ_4) will improve the energy dissipation by the highest amount when extended by Δt. Therefore, the first time quantum is assigned to task τ_4. In this manner, time quanta are iteratively distributed among the tasks until deadlines prevent further scalings. This optimisation process is illustrated through Table 3.3.   The table shows the evolving energy-gradients for each of the five tasks and highlights the task to which a time quantum $\Delta t = 0.01ms$ is assigned, that is, the task with the highest energy-gradient. The first row holds the energy-gradients that have been calculated just above. As mentioned, task τ_4 gains most out of an extension by $\Delta t = 0.01ms$. Hence, the time quantum is assigned to task τ_4, reducing its energy consumption by $1.130\mu J$ from $15\mu J$ to $13.870\mu J$. Having extended this tasks, its energy-gradient needs to be recalculated as $\Delta E_4 = E_4(0.16) - E_4(0.17) = 13.870\mu J - 12.905\mu J = 0.965\mu J$. The updated energy-gradient is shown in the second row of Table 3.3 (iteration 2). No time quanta have been distributed to the tasks τ_0, τ_1, τ_2, and τ_3, thus, their energy-gradients remain unchanged. Considering the second iteration, task τ_4 still achieves the highest energy saving when extended by $\Delta t = 0.01ms$, with $\Delta E_4 = 0.965\mu J$. Therefore, task τ_4 is extended by a second Δt and its energy-gradient is updated to 0.833. In the third iteration task τ_0 gains most of an extension and thus is extended. This iterative extension of tasks is repeated until no further extensions are possible without violations of task deadlines. This can be seen, for example, in the 15th iteration. Here the previous task extensions result in a situation where the task τ_4 just finishes in time, i.e., before its deadline is exceeded. Any further delay of task τ_4 or of any task that influences the finishing time of task τ_4 (i.e. the τ_0, τ_1, τ_2) would render the schedule infeasible. For this reason the tasks τ_0, τ_1, τ_2, and τ_4 are removed from the list (Table 3.3). Similarly, the last remaining task τ_3 reaches its deadline after the 16th iteration. At this point, no further scaling is possible and the distribution of slack is aborted. Note, although time quanta of $0.01ms$ are distributed 16 times, only the available slack of $0.1ms$ is exploited. This is due to the fact that some task extensions require slack from both PEs (because of task dependencies), while others are restricted to the slack of one PE. The last row of Table 3.3 shows how much the individual tasks have been extended. Accordingly, the new execution times are given as follows: $t_0 = 0.15ms + 0.04ms = 0.19ms$, $t_1 = 0.3ms + 0ms = 0.3ms$, $t_2 = 0.75ms + 0ms = 0.75ms$, $t_3 = 0.15ms + 0.06ms = 0.21ms$, and $t_4 = 0.15ms + 0.06ms = 0.21ms$. These extended execution times allow to lower the supply voltages during the execution of tasks τ_0, τ_3, and τ_4 to $4.349V$, $2.717V$, and $4.113V$ (using Equation 3.2), respectively, while the tasks τ_1 and τ_2 are ran at nominal supply voltage. Thus, the power dissipations of the tasks are: $P_{Vdd}(\tau_0) = 50.77mW$, $P_{Vdd}(\tau_1) = 20mW$, $P_{Vdd}(\tau_2) = 15mW$, $P_{Vdd}(\tau_3) = 38.74mW$, and $P_{Vdd}(\tau_4) = 48.33mW$. This results in a task energy dissipation of $E = 50.77mW \cdot 0.19ms + 20mW \cdot 0.30ms + 15mW \cdot 0.75ms + 38.74mW \cdot 0.21ms + 48.33mW \cdot 0.21ms = 45.18\mu J$. Adding the communication energy, the total energy consumption is given by $45.18\mu J + (5mW \cdot 0.05ms + 5mW \cdot 0.10ms) = 45.93\mu J$. This represents an energy reduction of 20.47%, which substantially higher compared to a reduction of 8.2% obtained with the power profile neglecting approach. Concluding, the power variations should be taken into account during the voltage selection in order to achieve further reductions in the energy dissipation. Many previous

DVS approaches for distributed system have considered the fixed power model, i.e., power variations were neglected. To overcome this limitation, the following section introduces a new voltage scaling technique capable of taking power variations into account with the aim to improve energy savings.

3.2   Algorithms for Dynamic Voltage Scaling

Having briefly outlined a new energy-gradient-based voltage scaling strategy in the previous section and shown the advantage of taking the power variations into account, this section introduces effective algorithms to implement this approach. Section 3.2.1 describes an novel energy-gradient voltage scaling, which is based on a new mapped-and-scheduled task graph structure (MSTG). In Section 3.2.2, this algorithm is enhanced towards discrete voltage selection. Section 3.2.3 analyses the computational complexity of the proposed voltage scaling algorithm.

3.2.1   Energy-Gradient-Based Voltage Scaling

The aim of the introduced DVS approach for distributed systems is to identify scaling voltages under the consideration of power variation effects. This is done for the scheduled and mapped system specification, such that the total dynamic energy dissipation is minimised. The presented approach assumes that no restrictions are placed on the scaling voltages, i.e., the technique targets variable-voltage systems (nearly continuous range of possible supply voltages) rather than multi-voltage systems (small and limited number of potential supply voltages). One particular aim of the voltage scaling technique is the energy estimation during the co-synthesis. Therefore, low optimisation times are of crucial importance, since the voltage scaling is executed in the innermost loop of an iterative co-synthesis process. One step towards this goal is the mapped-and-scheduled task graph structure, which will be introduced as part of the following voltage scaling algorithm. The overall voltage scaling algorithm is summarised in Figure 3.4, which is outlined next.

The input to the algorithm consists of the task graph specification, which has been mapped and scheduled beforehand onto a given system architecture. Furthermore, execution times and power dissipations are part of the architectural information, which also includes other necessary component properties, like the nominal supply voltage V_{max}, the threshold voltages V_t, etc. The minimal extension time Δt_{min} denotes the minimal time quantum to be distributed in each step of the algorithm. It is defined in order to speed up the determination of the voltage selection by preventing insignificant small extensions, which would lead to trivial power reductions.

To allow a fast and correct extension of task executions, which might influence other tasks and communications of the system due to interdependen-

---

Algorithm: *PV-DVS*

Input: - task graph $G_S(\mathcal{T}, \mathcal{C})$
  - mapping
  - schedule
  - architectural information
  - minimum extension time Δt_{min}
Output: - energy optimised voltages $V_{dd}(\tau)$
  - dissipated dynamic energy E

01: MSTG_TRANSFORM   //Generate MSTG from G_S
02: $\mathbb{Q}_E \leftarrow \varnothing$
03: **forall** $(\tau \in \mathcal{T}_d)$ $\{\Delta t_d(\tau) := t_d(\tau) - (t_S(\tau) + t_{exe}(\tau))\}$
04: **forall** $(\tau \in \mathcal{T})$ $\{$calculate $t_\epsilon\}$
05: **forall** $(\tau \in \mathcal{T})$ $\{$**if** $l_\epsilon \geq \Delta t_{min}$ **then** $\mathbb{Q}_E := \mathbb{Q}_E + \tau\}$
06: $\Delta t = \frac{\min t_\epsilon}{|\mathbb{Q}_E|}$, **if** $\Delta t < \Delta t_{min}$ **then** $\Delta t = \Delta t_{min}$
07: **for all** $(\tau \in \mathbb{Q}_E)$ $\{$calculate $\Delta E(\tau)\}$
08: reorder $\mathbb{Q}_E$ in decreasing order of ΔE
09: **while** $(\mathbb{Q}_E \neq \varnothing)$ $\{$
10:        select first task $\tau_{\Delta Emax} \in \mathbb{Q}_E$
11:        $t_{exe}(\tau_{\Delta Emax}) := t_{exe}(\tau_{\Delta Emax}) + \Delta t$
12:        update $E_{\tau_{\Delta Emax}}$
13:        **forall** $(\tau \in \mathcal{T})$ $\{$update t_S, t_E and $t_\epsilon\}$
14:        **forall** $(\tau \in \mathbb{Q}_E)$ $\{$**if** $(t_\epsilon(\tau) < \Delta t_{min}) \vee (V_{dd}(\tau) \leq V_t(\tau))$
                          then $\mathbb{Q}_E := \mathbb{Q}_E - \tau\}$
15:        $\Delta t = \frac{\min t_\epsilon}{|\mathbb{Q}_E|}$, **if** $\Delta t < \Delta t_{min}$ **then** $\Delta t = \Delta t_{min}$
16:        **forall** $(\tau \in \mathbb{Q}_E)$ $\{$update $\Delta E(\tau)\}$
17:        reorder $\mathbb{Q}_E$ in decreasing order of ΔE
18: $\}$
19: delete MSTG
20: **return** E_Σ, and $V_{dd}(\tau), \forall (\tau \in \mathcal{T})$

Figure 3.4. Pseudo code of the proposed heuristic (**PV-DVS**) algorithm

cies, it is beneficial to capture the schedule and mapping information into the task graph (line 01 in Figure 3.4). This can be achieved by generating a mapped-and-scheduled task graph (MSTG) structure, which is a transformed copy of the initial task graph. Figure 3.5 illustrates this transformation. The transformation consists of two steps, also given in the pseudo code of the MSTG_TRANSFORM algorithm (Figure 3.6): Firstly, all communications (edges) that are mapped to communication links are replaced by pseudo communication nodes and appropriate edges, thereby preserving the specified func-

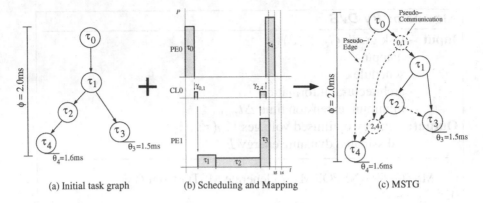

(a) Initial task graph   (b) Scheduling and Mapping   (c) MSTG

Figure 3.5. Capturing the mapping and schedule information into the task graph by using pseudo edges and communication task

tionality. Secondly, all nodes mapped to a certain PE or CL are traversed in chronological order of execution and linked by pseudo-edges, if an edge does not already exist. In this way, the schedule and mapping are inherited into the task graph, and the influence of a task extension can be easily propagated through the system schedule by traversing the MSTG in a breadth-first order. Consider, for instance, the extension of task τ_2. The initial task graph (Figure 3.5(a)) does not reveal if the extension of task τ_2 has an influence on any other task or communication, except on task τ_4. However, the scheduling and mapping shown in Figure 3.5(b) indicates that an extension of task τ_2 will influence the subsequent task τ_3. The MSTG shown Figure 3.5(c) captures this dependency through an pseudo edge between the tasks τ_2 and τ_3. The main advantage of this tactic lies within the linear time complexity of the breadth-first search algorithm [63], which is leveraged to update the start and end times of all influenced activities (tasks as well as communications), due to voltage scaling.

After the initial task graph has been transformed into a MSTG, the DVS continues with line 02 (Figure 3.4), where a priority queue $\mathbb{Q}_E$ is initialised. In order to identify all extendable tasks, the algorithm calculates the available slack times Δt_d of each hard deadline task $\mathcal{T}_d$ (line 03). This slack is given by the difference between task deadline $t_d(\tau)$ and the task finishing time $t_S(\tau) + t_{exe}(\tau)$, where t_S and t_{exe} denote the task start and execution time, respectively. The algorithm then calculates the available slack times t_ϵ of all tasks, taking into account the interrelation between tasks and communications (line 04). For this purpose an inverse breadth-first search algorithm is used to visit all nodes of the MSTG to inherit the slack time of influenced tasks. If a tasks has several successors, the smallest slack is inherited to ensure that no deadlines will be exceeded during the voltage scaling. In line 05, the algorithm includes tasks with an available slack time t_ϵ greater or equal than the minimal extension time

Δt_{min} into the priority queue $\mathbb{Q}_E$. In this way, tasks with negligible small or no extension possibility are excluded from the scaling process. The initial extension time Δt is calculated by dividing the smallest slack time among the extendable task ($\min t_\epsilon$) by the number of extendable tasks $|\mathbb{Q}_E|$ (line 06). This extension time, however, should not be smaller than the minimal extension time Δt_{min} (the selection of an appropriate Δt_{min} will be discussed in Section 3.3). It is now possible to calculate the energy-gradients of all extendable tasks, that is, the tasks in the queue $\mathbb{Q}_E$, using the Equations (3.1), (3.2), and (3.4), as indicated in line 07. In line 08, the priority queue $\mathbb{Q}_E$ is reordered in decreasing order of the energy-gradients.

Now the distribution of slack starts. The algorithm iterates the steps between lines 09 and 18 until no extendable tasks are left in the priority queue $\mathbb{Q}_E$. In each of these iterations the algorithm picks the first element from the priority queue, the task which leads to the highest energy reduction (line 10). This task

---

Algorithm: *MSTG_TRANSFORM*

Input: - task graph $G_S(\mathcal{T}, \mathcal{C})$
           - scheduling and mapping information
Output:- mapped and scheduled task graph (MSTG)

```
01: forall (γ ∈ C) {    //insert communication tasks
02:        if (γ is mapped to any link L) {
03:            insert new communication task τ_C
04:            insert edge γ_t from task source(γ) to task τ_C
05:            insert edge γ_f from task τ_C to task sink(γ)
06:            remove γ
07:        }
08: }
09: forall (κ ∈ K) {    //insert pseudo edges
10:        predecessor task τ_P ← NIL
11:        traverse all task τ_κ mapped to κ
           in chronological order {
12:            if (τ_P ≠ NIL) ∧ (γ_{P→κ} = NIL)
13:                insert edge γ_{P→κ} from τ_P to τ_κ
14:            else
15:                τ_P = τ_κ
16:        }
17: }
```

Figure 3.6. Pseudo code of task graph to mapped-and-scheduled task graph transformation

is then extended by Δt and the energy dissipation value is updated (lines 11
and 12) according to Equations (2.8) and (2.10). In line 13, the extension is
propagated through the MSTG, since successor tasks might have been affected
by the extension in terms of start time t_S, end time $t_E = t_S + t_{exe}$, and available
slack time t_ϵ. This propagation is carried out using a breadth-first search to visit
all successors of the extended task, i.e., tasks and communications that might
have been affected by the extension. In the next step (line 14), inextensible tasks
are removed from the extendable task queue $\mathbb{Q}_E$ if their available slack t_ϵ is
smaller than the minimal extension time Δt_{min}, or their scaled supply voltage
V_{dd} is smaller or equal to the threshold voltage V_t. Taking into account the tasks
in the new extendable queue, the time quantum Δt is recalculated (line 15), and
based on this value, the energy-gradients ΔE are updated (line 16). The priority
queue $\mathbb{Q}_E$ is reordered according to the new energy-gradients (line 17). At this
point, the algorithm either invokes a new iteration or terminates, based on the
state of the extendable task queue. On algorithm completion, the MSTG copy
is deleted and the scaling voltages for each task execution as well as the total
dynamic energy dissipations are returned (lines 19 and 20).

3.2.2   Discrete Voltages

The algorithm, as described above, produces scaling voltages under the assumption that variable-voltage PEs are available that support continuous voltage
scaling. However, it is possible to adapt the generated scaling voltages towards
multi-voltage PEs, which are able to run at a restricted number of predefined
voltages, as for example the case for the processor presented by Burd *et al.* [34,
36]. This processor has the ability to run at 15 different operational frequency
(5 to 80 MHz) and voltage settings. It has been shown in [80] that the two
discrete supply voltages V_{d1} and V_{d2}, $V_{d1} < V_{dd} < V_{d2}$, around the continuous
selected voltage V_{dd} are the ones which can be used to minimise the energy
dissipation, under the assumption that the time overhead for switching between
different voltages can be neglected. (Although, these two voltages do not necessarily optimally minimise the energy consumption [162].) Thus, the proposed
approach can be used for voltage selection on multi-voltage PEs. Given a task
τ with execution time t_{exe} at the continuous selected voltage V_{dd}, then, in order
to achieve minimal energy consumption, the same task τ will execute on the
multiple voltage PE for t_{d1} time units at the supply voltage V_{d1} and for t_{d2} time
units at supply voltage V_{d2}, where

$$t_{exe} = t_{d1} + t_{d2} \tag{3.5}$$

$$t_{d1} = t_{exe} \cdot \frac{V_{d1} \cdot (V_{dd} - V_t)^2}{(V_{d1} - V_t)^2 \cdot V_{dd}} \cdot \frac{\frac{V_{dd}}{(V_{dd}-V_t)^2} - \frac{V_{d2}}{(V_{d2}-V_t)^2}}{\frac{V_{d1}}{(V_{d1}-V_t)^2} - \frac{V_{d2}}{(V_{d2}-V_t)^2}}. \tag{3.6}$$

Nevertheless, since the task execution times are given by

$$t_i = NC_i \cdot f_i \qquad (3.7)$$

where NC_i denotes the number of clock cycles executed at frequency f_i, it is clear that the calculated execution times might have to be adjusted because clock cycles are given as *integer* values. This is because a task execution is performed on a cycle-by-cycle base. Therefore, the calculated execution times (t_{d1} and t_{d2}) are adjusted in the following way: The floor integer number of clock cycle NC_{d1} for the execution at low voltage V_{d1} is calculated as:

$$NC_{d1} = \lfloor t_{d1} \cdot f_{d1} \rfloor \qquad (3.8)$$

While the number of clock cycles NC_{d2} executed at the high voltage V_{d1} is given by:

$$NC_{d2} = NC_{tot} - NC_{d1} \qquad (3.9)$$

where NC_{tot} is the total number of clock cycles a task requires for its execution. Thereby it is insured that the overall task execution time t_{exe} is not exceeded. The adjusted execution times at voltages V_{d1} and V_{d2} are then given by,

$$t_{d1} = NC_{d1}/f_{d1} \qquad (3.10)$$

$$t_{d2} = NC_{d2}/f_{d2} \qquad (3.11)$$

3.2.3 Algorithm Complexity

As mentioned in Section 3.2.1, the voltage scaling algorithm is intended to be used within the innermost loop of the co-synthesis where scheduling and mapping are iteratively optimised. Therefore, a moderate computational complexity is desirable in order to allow a thorough exploration of the design space in reasonable amounts of time. The complexity of the proposed PV-DVS algorithm can be derived as follows. The initial task graph can be copied and transformed into a MSTG in linear time. The WHILE loop (line 09 in Figure 3.4) is executed in the worst case $n \cdot m$ times, where n is the number of nodes $|\mathcal{T}|$ in the graph, since all tasks might be extendable. However, depending on Δt_{min} and Δt, tasks might be extended more than once, and m, for the worst case, is the maximum number of such extensions. The inner part of the WHILE loop shows the following complexities: The propagation of extensions takes $n + e$ in the worst case ($e = |\mathcal{C}|$ is the number of edges in the graph), since all nodes and edges might have to be visited by the breadth-first search (line 13). Removing inextensible tasks, again, might take n steps. At most n steps are needed to determine the new extension time Δt. And finally, updating the extendable queue takes n operations (the queue is implemented as Fibonacci heap [59]). All other calculations inside the WHILE loop are executed in constant time.

Therefore, the time complexity of the proposed PV-DVS algorithm is given as $\mathcal{O}(n \cdot m(4n + e))$. It should be noted that the extendable task queue $\mathbb{Q}_E$ is progressively reduced from length n to 0. The reduction is not uniform since it might occur that suddenly (at the same time) many tasks become inextensible and are excluded from the queue. This, additionally, indicates that the complexity derived above is valid for the worst case. Furthermore, to account for DVS-PEs which run at discrete voltages, the suitable supply voltage have to been derived from the continues voltage. This can be done in linear time. In summary, the overall computational complexity (including MSTG generation, PV-DVS, and discrete voltage calculation) is quadratic in the number of tasks and communications.

3.3   Experimental Results: Energy-Gradient based Dynamic Voltage Scaling

To demonstrate the efficiency and the applicability of the proposed generalised DVS technique (considering the power variation model) in reducing the energy dissipation of heterogeneous distributed systems containing power-managed PEs, numerous experiments have been carried out. The DVS algorithm outlined in this chapter has been implemented using C++ on a Pentium-III/750MHz Linux PC with 128MB RAM. The used benchmark examples consist of 7 task graphs taken from previously published literature [20, 78] and 28 task graphs automatically generated using TGFF [48]. These benchmarks are used to cover a wide spectrum of application diversity. The complexity of these task graph examples varies between 8 to 100 tasks with 7 to 151 edges. The amount of processing elements (PEs) and communication links (CLs) in the technology libraries varies between 4 and 16. The benchmarks used in this chapter are grouped into three major sets:

(a) Our TGFF generated task graphs (tgff1-tgff25, tgff4_t, tgff4_fixed), and tgff17_m are mapped onto heterogeneous architectures containing power-managed DVS-PEs and standard PEs without DVS. Thus, these examples show varying power characteristics and component properties. The benchmark examples tgff4_t and tgff4_fixed are identical to tgff4 with slight modifications; tgff4_t denotes a task graph alternative with a critical tight deadline, while tgff4_fixed uses only DVS-PEs with a fixed power dissipation. Similarly, the example tgff17_m is a slightly modified version of tgff17 with a different hardware architecture.

(b) The examples of Hou and Wolf [78] are two hypothetical task graphs (Hou and Hou_clustered). The task graph Hou_clustered represents the same functionality as Hou but the task graph is collapsed from 20 to 8 tasks. Since their technology library does not contain any DVS-enabled PEs, the given PEs are extended towards DVS capability (with $V_t = 0.8V$ and

$V_{max} = 3.3V$). These examples also show different power dissipations (power variations) among the tasks, unlike the first benchmarks set.

(c) The applications used by Bambha *et al.* [20] consist of two differently implemented fast Fourier transformations (fft1 and fft3), a Karplus-Strong music synthesis algorithm (Karp10), a quadrature mirror filter bank (qmf4), and a measurement application (meas). These real-life benchmarks of moderate size (12–28 tasks) use architectures composed of 2 to 6 identical DVS-PEs, assuming constant power dissipation. Supply voltages are between 0.8 and 7 volts. The throughput constraints and initial average power consumptions are calculated at a reference voltage of 5 volts.

The following experimental results are split into two sections. The first set of experiments concentrates on the influence of the power variation effect (Section 3.3.1), while the second demonstrate the importance of an appropriate selection of the minimal extension time (Section 3.3.2).

3.3.1 Power Variations

To demonstrate the influence of power variations on the efficiency of DVS, the PV-DVS algorithm (Section 3.2) is compared to a power neglecting approach, i.e., a DVS approach that makes use of the fixed power model. The power neglecting approach (in the following referred to as EVEN-DVS) is based on the intuitive idea to distribute available slack time *evenly* among the processing elements, somewhat similar to the slack distribution idea used in [99].

Results of Benchmarks (a) and (b):
Table 3.4 shows a comparison between the EVEN-DVS (fixed power model) and the PV-DVS approach (power variation model) using the tgff1-tgff25 benchmarks as well as the two examples from Hou *et al.* [78] (Hou and Hou_clustered). In order to judge the complexity of the individual benchmark examples, the table gives the number of nodes and edges in the task graphs. The comparison between the two DVS approaches is carried out with respect to the energy dissipation when no DVS is employed, i.e., when the tasks are executed at nominal supply voltage (highest energy consumption). Consider, for example, benchmark tgff17, which consist of 29 tasks and 56 communications between tasks. The unscaled execution (NO-DVS) of the application dissipates an energy of $23459\mu J$. Using an even distribution of slack time (EVEN-DVS) this power consumption can be reduce to $20396.41\mu J$, a reduction of 13.06%. However, considering the power variations by using the PV-DVS algorithm it is possible to further reduced the energy to $18334.01\mu J$, when compared to the nominal energy dissipation a reduction of 21.85%. For all examples shown in Table 3.4 it is assumed that the mapping and schedule have been pre-determined, using a static mapping and a schedule generated by a mobility-

Example	No. of tasks / edges	NO-DVS (nominal) Energy (μJ)	EVEN-DVS (fixed power model) Energy (μJ)	Reduction (%)	PV-DVS (power variation model) Energy (μJ)	Reduction (%)
tgff1*	8 / 9	355	193.49	45.50	112.87	68.21
tgff2	26 / 43	743224	722412.15	2.80	683954.54	7.97
tgff3	40 / 77	554779	410653.67	25.98	267651.03	51.76
tgff4	20 / 33	431631	402904.08	6.66	375914.03	12.91
tgff4_t	20 / 33	431631	412854.36	4.35	397201.93	7.98
tgff4_fixed	20 / 33	176723	142986.91	19.09	124905.61	29.32
tgff5	40 / 77	4187382	3963647.60	5.34	3767450.25	10.03
tgff6	20 / 26	1419124	1401605.68	1.23	1396445.06	1.60
tgff7	20 / 27	2548751	2289878.34	10.16	1951579.52	23.43
tgff8	18 / 26	1913519	1774151.42	7.28	1668485.33	12.81
tgff9*	16 / 15	996590	974159.01	2.25	918048.34	7.88
tgff10	16 / 21	69352	51263.60	26.08	46483.97	32.97
tgff11	30 / 29	4349627	4293736.56	1.28	4263279.98	1.99
tgff12	36 / 50	2316431	2243710.55	3.14	2212111.25	4.50
tgff13	37 / 36	2912660	2425431.77	16.73	2333338.86	19.89
tgff14	24 / 33	15532	13546.62	12.78	12479.41	19.65
tgff15	40 / 63	62607	62078.93	0.84	60334.62	3.63
tgff16	31 / 56	3494478	2913341.14	16.63	2518711.99	27.92
tgff17	29 / 56	23459	20396.41	13.06	18334.01	21.85
tgff17_m	29 / 56	153120	134988.02	11.84	113889.21	25.62
tgff18	29 / 57	1277704	1196851	6.38	665969	47.88
tgff19	14 / 19	5939	4713.59	20.63	4395.37	25.99
tgff20*	19 / 25	77673	48334.30	37.77	40280.98	48.14
tgff21	70 / 99	3177705	3175497.22	0.07	2658534.22	16.34
tgff22	100 / 135	5821498	5036657.40	13.48	4445545.63	23.64
tgff23*	84 / 151	11567283	10791880.89	6.70	10133912.03	12.39
tgff24	80 / 112	5352217	5349024.86	0.06	5238478.58	2.13
tgff25	49 / 92	5735038	5648816.00	1.50	5502681.64	4.05
Hou*	20 / 29	12265	10337.05	15.72	7474.55	39.06
Hou_clustered*	8 / 7	14546	12661.78	12.95	10270.32	29.39

*Components used for these examples consists of DVS-PEs only

Table 3.4. Comparison of the presented PV-DVS optimisation with the fixed power model using EVEN-DVS approach

based list scheduling technique [158]. Thus, the energy reductions are solely achieved through voltage scaling. As expected, both the EVEN-DVS and the PV-DVS technique reduced the energy dissipation of the systems in all cases (Column 6 and 9). It can be observed that the proposed DVS heuristic (PV-DVS) was able to further reduce the energy dissipation of all examples, when compared to EVEN-DVS. Due to the particular implementation of the DVS algorithm, which distributes slack evenly among the PEs (EVEN-DVS), also slack is allocated on non-DVS-PEs. Therefore, the higher energy reductions of the proposed DVS algorithm are due to two facts. Firstly, the EVEN-DVS allocates slack time on non-DVS-PEs. These times, of course, cannot be exploited

Example	No. of Nodes/ Edges	NO-DVS Energy Dissip.	PV-DVS (power variation model)		
			Energy Dissip.	*CPU time (s)*	*Reduction (%)*
fft1	28/32	29600	18154	0.12	38.67
fft3	28/32	48000	36772	0.13	23.39
karp10	21/20	59400	47737	0.12	19.63
meas	12/12	28300	25732	0.10	9.07
qmf4	14/21	16000	12740	0.11	20.38

Table 3.5. PV-DVS results using the benchmarks of Bambha *et al.* [20]

to lower the power consumption. Secondly, the proposed DVS technique considers the power variations during the voltage scaling. This leads to better energy reductions (see Motivational Example 1, Section 3.1). To distinguish between both effects, architectures that consists of DVS-PEs only have been indicated in Table 3.4. In these examples, the higher energy reduction is solely achieved by taking the power profile into account. The remaining examples achieve increased energy efficiency due to both effects. It should be noted that it is hard to judge the achievable energy savings in a straightforward way, since the savings depend on task interdependencies as well as on the individual power dissipations of the given tasks.

In order to get an idea about the voltages scaled schedules consider Figure 3.7. This figure shows the task execution of benchmark tgff17_m for three different situations. Firstly, Figure 3.7(a) depicts the schedule at nominal supply voltage, i.e., tasks are executed as fast as possible ($f = f_{max}$ and $V_{dd} = V_{max}$), hence, consuming the maximal energy. Secondly, Figure 3.7(b) corresponds to an even distribution of slack time. And finally, Figure 3.7(c) illustrates the scaled execution based on the proposed PV-DVS algorithm. In this particular benchmark only one processing element (PE2) is DVS enabled, therefore, the execution properties of tasks mapped onto PE0 and PE1 are static.

Results of Benchmark (c):
The next experiments are concerned with the benchmark set used by Bambha *et al.* [20]. Since they use a different communication model (contention, requests for the bus, etc.), the throughput constraints had to be re-calculated. Therefore, a direct comparison between the results reported in [20] and the results presented here is not valid. Nevertheless, the re-calculation of the throughput was carried out for the same task mapping and execution order as in [20], which is based on a dynamic level scheduling approach [138]. The results of these five examples, obtained using the PV-DVS method, are given in Table 3.5. It can be observed that in all cases the energy was decreased, with reductions between 9.07% to

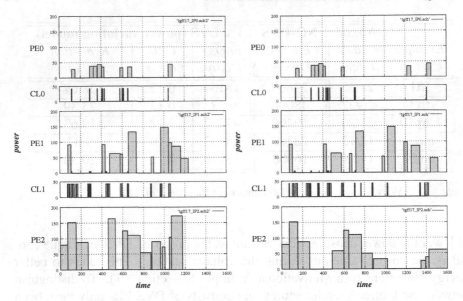

(a) Task execution at nominal supply voltage (**no** DVS is exploited)

(b) Task execution at scaled supply voltages using the EVEN-DVS approach (fixed power model)

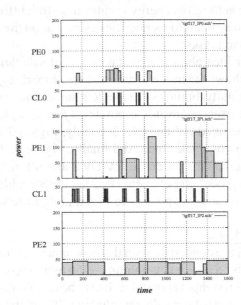

(c) Scaled task execution using the proposed PV-DVS algortihm (power variation model)

Figure 3.7.   Three identical execution orders of the `tgff17_m` benchmark: (a) unscaled execution at nominal supply voltage (NO-DVS), (b) using the EVEN-DVS, and (c) the PV-DVS approach

38.67%. Furthermore, the highly serialised structure of meas allowed us to calculate the theoretically optimal voltage schedule for this example. Using the optimal supply voltages results in 13% energy saving. The PV-DVS algorithm achieved for the same example a reduction of 9.07%, which is only 3.93% higher than the theoretically optimal solution. The reason for this for this is the greediness of the PV-DVS algorithm which always slacks the task which achieves the highest possible energy savings on a single PE; however, scaling this single task might require slack on several different PEs.

The presented results assume that computation and voltage scaling can be carried out concurrently, as the case of the processor introduced in [34]. Further, the time overhead needed by the processor to switch between two supply voltages is neglected since the used tasks are considered to be of coarse granularity. Therefore, the switching can be considered to be only a small fraction of the total task execution time (for real-world DVS processors this is in the range of 10–$70\mu s$ for a full transition from the highest to the lowest supply voltage and vice versa [34]). However, in the case of fine-grained task this overhead might influence the optimisation and should therefore be considered. Techniques which consider the scaling overhead have been recently proposed [17, 77, 106, 162].

3.3.2 Minimal Extension Time

To give insight into the dependencies between the computation effort, solution quality, and the minimal extension time Δt_{min} (see also the complexity analysis in Section 3.2.3), two additional experiments were conducted. In order to achieve accurate results, especially for the measurement of the optimisation times, the experiments are carried out using two large task graphs with 80 (tgff23) and 400 (large1) tasks. Figure 3.8 illustrates the dependency between the minimal extension time Δt_{min} and the solution quality (given as reduced energy E over nominal energy E_0). It can be observed that no energy reduction can be achieved until Δt_{min} is smaller than the largest slack available in the activity schedule (2364 for tgff23 and 29581 for large1, see Figure 3.8). Clearly, if the algorithm must distribute time quanta bigger than any available slack, it cannot perform any voltage reduction, and therefore $E/E_0 = 1$. But energy reductions can be achieved by decreasing Δt_{min} to a value below the biggest slack that is present on a DVS-PE. In the case of example tgff23, the energy consumption approaches 87% of the nominal energy when reducing Δt_{min} below 2364. Nevertheless, it is not desirable to decrease the minimal extension time too much, since the additional reductions become insignificant (the curves level out) and the unnecessary extension will only increase the computational time of the optimisation.

The dependency between minimal extension time Δt_{min} and optimisation time of the DVS algorithm is illustrated in Figure 3.9, using a double loga-

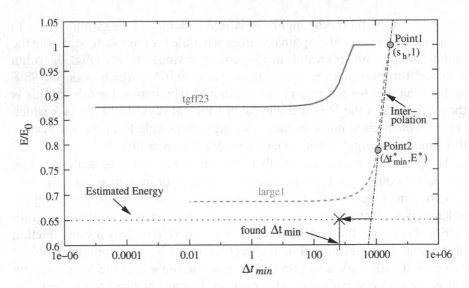

Figure 3.8.   Energy reduction quality dependent on minimal extension time Δt_{min}

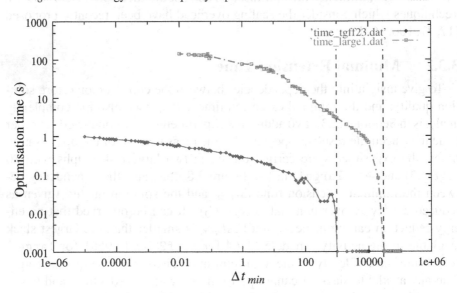

Figure 3.9.   Execution time dependent on minimal extension time Δt_{min}

rithmic scale. It can be seen that with decreasing Δt_{min} the optimisation time increases. Similar to the observations given before, the algorithm cannot start any optimisation before Δt_{min} is smaller than the biggest slack given in the schedule (indicated by the vertical dashed lines in Figure 3.9). It is therefore important to find a good value for Δt_{min}, which trades off between solution quality and optimisation time. Finally, Figure 3.10 shows the energy reduction

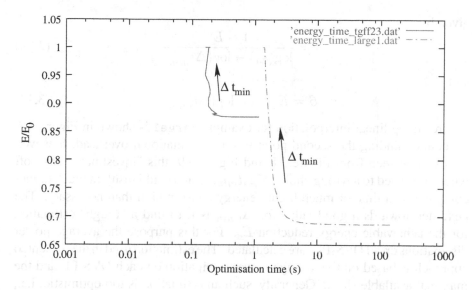

Figure 3.10. Energy reduction quality dependent on execution time

dependent on the optimisation time. As the plot illustrates, with an increasing Δt_{min} the optimisation time decrease while the energy dissipation increases due to a less accurate calculation of the scaled supply voltages.

These experiment have shown that a well chosen Δt_{min} is essential for a fast and accurate energy estimation using the **PV-DVS** algorithm. Accordingly, a good choice is important to speed up the voltage scaling process. The following describes a heuristic method that has been used in the presented experiments to find an appropriate Δt_{min} setting for each solution candidate. It is based on the observation that for all conducted benchmarks experiments the characteristics shown in Figure 3.8 and Figure 3.9 hold. With reference to Figure 3.8, the nearly linear (in a semi-logarithmic scale) energy drop, after decreasing Δt_{min} below the highest DVS slack, is interpolated by a logarithmic function (indicated as "Interpolation" in Figure 3.8):

$$y = \alpha \cdot \log x + \beta \qquad (3.12)$$

where the constants α and β are calculated using two initial points in the quasi linear part of the graph. The first point (Point 1) corresponds to the highest available slack s_h on any of the DVS-PEs, hence, matches the nominal energy dissipation. This point can be found in linear time and is indicated in Figure 3.8. To establish the second point needed for the interpolation, the DVS algorithm is run with a Δt_{min}° three times smaller than the highest DVS slack to find its corresponding reduced energy dissipation E°. For all used examples this was still in the steeply dropping part of the graph. The second point shown in Figure 3.8 was found in this way. Using both points, the constants α and β are given by:

$$\alpha = \frac{1 - E^\circ}{\log(s_h) - \log(\Delta t^\circ_{min})} \tag{3.13}$$

and

$$\beta = E^\circ - \alpha \cdot \log(\Delta t^\circ_{min}) \tag{3.14}$$

The resulting linear interpolation for example `large1` is shown in Figure 3.8. Of course, finding the second point has a computational overhead, however, as it can be seen from Figure 3.8 and Figure 3.9, this "investment" pays off when compared to a wrong choice of Δt_{min}, which could result in much higher computational time or much higher energy consumption than necessary. The next step towards a good value for Δt_{min} is it to find a "rough" estimation for the achievable energy reduction E_e. For this purpose the average power dissipations on all DVS-PEs are calculated. The estimation used in the presented approach is based on the average power dissipation on each DVS-PE and the maximal available slack. Generally, such an estimation is too optimistic, i.e., it will be lower than the in reality achievable energy dissipation. An estimated energy dissipation for example `large1` is indicated in Figure 3.8. The minimal extension time Δt_{min} could be set to the intersection of the energy estimation and the interpolated energy drop. However, in the proposed heuristic the value of Δt_{min} is set one order of magnitude lower (as indicated by the cross in Figure 3.8). This is done to account for the fact that an energy estimation close to the real achievable energy reduction would be approximately one order of magnitude away from a good Δt_{min}. On the other hand, if the energy estimation would be far below the real achievable energy reduction, the calculated Δt_{min} would become unnecessary small. Therefore, no Δt_{min} smaller than 2.5 orders of magnitude (compared to the maximal DVS slack) is allowed. This is based on the observation that all used benchmarks show similar characteristics, and that an ideal Δt_{min} can mostly be found at maximal 2.5 orders of magnitude from the maximal DVS slack. It is fair to say, these extreme situations appear rarely during the optimisations process.

3.4   Concluding Remarks

This chapter has introduced a new energy minimisation technique for dynamic voltage scaling in distributed heterogeneous systems. The proposed technique is based on an energy-gradient based heuristic that takes into account variations in the power profile of DVS processing elements as well as the heterogeneity of the scalable components. In order to keep the computational complexity low, a novel mapped-and-scheduled task graph (MSTG) structure was proposed, which allows the propagation of task extensions in linear time. The selection of the minimal extension time was addressed through a heuristic approach that leverage observations taken from extensive experiments. Experimental results have reinforced the argument that power variations should be taken into account, in order to achieve higher energy saving compared to approaches that consider a fixed power model. Although the introduced PV-DVS algorithm has an overall quadratic computational complexity (while the fixed-power-model DVS can be applied in linear time), this complexity is low enough to scale system specifications of realistic size (10–100 tasks) in small amount of run-time.

Chapter 4

OPTIMISATION OF MAPPING AND SCHEDULING FOR DYNAMIC VOLTAGE SCALING

Chapter 3 has shown how the energy consumption of distributed embedded systems can be minimised using a dynamic voltage scaling (DVS) technique that accounts for the power variations (PV) of tasks. This PV-DVS technique was applied to applications that had been statically mapped and scheduled beforehand. Nevertheless, the efficiency with which DVS can be applied depends significantly on the available idle and slack times within the system schedule. Clearly, application mapping (Section 1.3, page 9) as well as activity scheduling (Section 1.3, page 11) affect considerably these idle and slack times. For this reason a co-synthesis, which incorporates mapping and scheduling, should tightly integrate the consideration of DVS, in order to find solutions that carefully increase the amount of available idle and slack times on the DVS-PEs. This chapter introduces new techniques and algorithms for scheduling and mapping in distributed systems, which aim, in addition to traditional design goals such as performance and area usage, at an optimised utilisation of the DVS-PEs, and hence the reduction of the energy consumption. The overall co-synthesis flow is outlined in Figure 4.1. As indicated, the introduced co-synthesis is split into two steps:

- Combined optimisation of scheduling and communication mapping

- Task mapping optimisation

The remainder of this chapter outlines this *two-step* co-synthesis process. The chapter is organised as follows. Section 4.1 concentrates on energy minimisation through schedule optimisation towards an effective utilisation of PV-DVS. Techniques and algorithms for the optimisation of task and communication mapping are introduced in Section 4.2. Section 4.3 demonstrates through the usage of a real-life example how the proposed techniques can be applied in

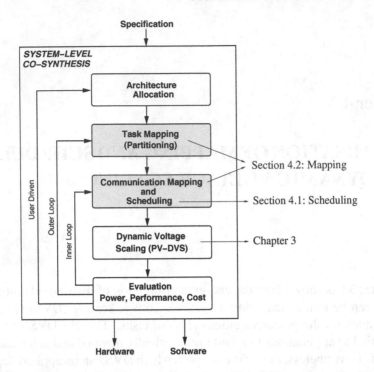

Figure 4.1. Co-synthesis flow for the optimisation of scheduling and mapping towards the utilisation of PV-DVS

order to optimise the system implementation (including hardware architecture) towards energy-efficiency. Finally, Section 4.4 gives a summary of this chapter and draws some conclusions.

4.1 Schedule Optimisation

This section is concerned with the scheduling problem of heterogeneous distributed systems that contain DVS-PEs. The scheduling of tasks and communications greatly influences how efficiently DVS can be exploited, due to the direct impact on the available slack times. In general, the more slack is available in the schedule, the higher will be the achievable energy savings by exploiting DVS. This, however, becomes much more complex and does not hold always for distributed systems under the consideration of the power variation model. In such a case, the available slack for high energy dissipating tasks should be considered more important than the slack of tasks consuming a minor amount of power.

This section first motivates the schedule optimisation under the power variation model in Section 4.1.1. This is followed by Section 4.1.2 which provides background information about scheduling techniques and genetic algorithms.

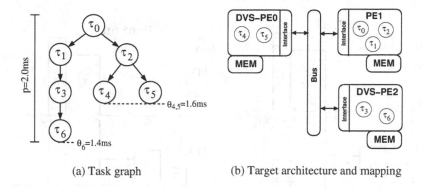

(a) Task graph   (b) Target architecture and mapping

Figure 4.2.   Specification and DVS-enabled architecture

Section 4.1.3 introduces a new scheduling algorithm for energy minimisation through PV-DVS. Finally, Section 4.1.4 presents experimental results.

4.1.1   Motivational Example: Scheduling

The purpose of this motivational example is to highlight the importance of taking power variations into account, while making energy-conscious scheduling decisions in the presence of DVS-PEs. Consider the system specification given as task graph in Figure 4.2(a). The seven tasks $\{\tau_0, ..., \tau_6\}$ are mapped onto the architecture as shown in Figure 4.2(b). The architecture consists of two DVS-PEs (PE0, PE2) and one non-DVS-PE (PE1), which are connected through a single bus. The nominal supply voltage V_{max} and the threshold voltage V_t of PE0 and PE2 are $V_{max} = 3.3V$ and $V_t = 0.8V$, respectively, while PE1 runs all tasks at V_{max} (it cannot be scaled). The task execution times t_{nom} and power dissipations P_{max} at nominal supply voltage are given in Table 4.1, which also shows the task mapping. For the sake of simplicity, the

Task	exe. time t_{nom} (ms)	power dis. P_{max} (mW)	mapped to
τ_0	0.3	10	PE1
τ_1	0.3	20	PE1
τ_2	0.4	15	PE1
τ_3	0.1	40	PE2
τ_4	0.4	70	PE0
τ_5	0.2	90	PE0
τ_6	0.3	20	PE2

Table 4.1.   Nominal execution times and power dissipations for the mapped tasks

communications are neglected when discussing this particular example. Figure 4.3(a) shows a possible schedule for the tasks executing at nominal supply

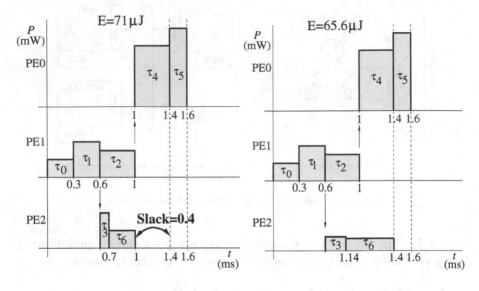

(a) Execution at nominal supply voltage V_{max} (b) Scaled execution with $V_{dd3} = 2.08V$ and
                                                          $V_{dd6} = 2.34V$

Figure 4.3.   A possible schedule *not* optimised for DVS

voltage, i.e., at maximal supply voltage and consequently the highest energy dissipation. According to the values given in Table 4.1, the energy dissipation of this schedule can be calculated as $E = \sum_{\tau \in \mathcal{T}}(P_{max}(\tau) \cdot t_{nom}(\tau)) = 0.3ms \cdot 10mW + 0.3ms \cdot 20mW + 0.4ms \cdot 15mW + 0.1ms \cdot 40mW + 0.4ms \cdot 70mW + 0.2ms \cdot 90mW + 0.3ms \cdot 20mW = 71\mu J$. Considering the task deadlines given in Figure 4.2(a), it can be observed from the schedule in Figure 4.3(a) that the tasks τ_3 and τ_6 are eligible for scaling, since τ_6 finishes execution after $1ms$, while its deadline is at $\theta_6 = 1.4ms$. Hence, leaving a slack of $0.4ms$. An extension of any other task cannot be tolerated, since task τ_5 finishes execution just on its deadline $\theta_5 = 1.6ms$. By scaling the schedule, using the proposed implementation of the **PV-DVS** technique (taking power variations into account) presented in Chapter 3, the voltage schedule shown in Figure 4.3(b) can be produced. In this scaled schedule the tasks τ_3 and τ_6 execute at $2.08V$ and $2.34V$, respectively. Using Equation (2.6) and considering that the switched load capacitance $N_C^\tau \cdot \alpha \cdot C_L$ is constant for a given task, the energy consumptions of the tasks τ_3 and τ_6 are reduced to

$$E_{V_{dd}}^{\tau_3} = E_{max}^{\tau_3} \cdot \frac{N_C^{\tau_3} \cdot \alpha \cdot C_L \cdot V_{dd}^2}{N_C^{\tau_3} \cdot \alpha \cdot C_L \cdot V_{max}^2} = 0.1ms \cdot 40mW \cdot \frac{(2.08V)^2}{(3.3V)^2} = 1.59\mu J$$

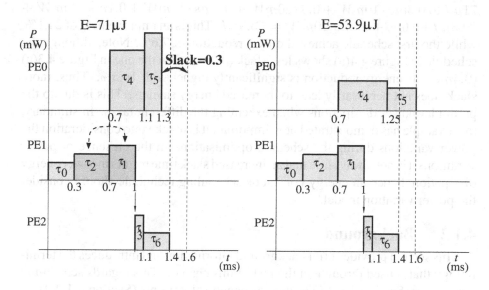

(a) Execution at nominal supply voltage V_{max} (b) Scaled execution with $V_{dd4} = 2.74V$ and $V_{dd5} = 2.41V$

Figure 4.4.   Schedule optimised for DVS considering the power variation model

and

$$E_{Vdd}^{\tau 6} = 0.3ms \cdot 20mW \cdot \frac{(2.34V)^2}{(3.3V)^2} = 3.01\mu J$$

Thereby, the total energy dissipation is given by $0.3ms \cdot 10mW + 0.3 \cdot 20mW + 0.4ms \cdot 15mW + 1.59\mu J + 0.4ms \cdot 70mW + 0.2ms \cdot 90mW + 3.01\mu J = 65.6\mu J$. This represents a reduction of 7.6%.

Now, consider a second feasible schedule at nominal supply voltage where the execution order of τ_1 and τ_2 has been swapped, as shown in Figure 4.4(a). This allows to start task τ_4 earlier, while the execution of task τ_3 is delayed. Since this re-scheduling does not influence the nominal execution times and power dissipations, the nominal energy consumption remains $E = 71\mu J$ (as in first nominal schedule of Figure 4.3(a)). Observing the schedule reveals that task τ_6 just finishes execution within its deadline constraint $\theta_6 = 1.4ms$, while task τ_5 terminates execution after $1.3ms$. This leaves a slack of $0.3ms$ until its deadline $\theta_5 = 1.6ms$ is met. Therefore, only the tasks τ_4 and τ_5 can be scaled. Applying the **PV-DVS** technique of Chapter 3 returns the scaling voltages $V_{dd4} = 2.74V$ and $V_{dd5} = 2.41V$ for the tasks τ_4 and τ_5, respectively. Executing the tasks at these voltages, the energy consumptions are reduced to $E_{Vdd}^{\tau 4} = 0.4ms \cdot 70mW \cdot (2.74V)^2/(3.3V)^2 = 19.3\mu J$ and $E_{Vdd}^{\tau 5} = 0.2ms \cdot 90mW \cdot (2.41V)^2/(3.3V)^2 = 9.6\mu J$. Hence, the total energy is reduced from

$71\mu J$ to $0.3ms \cdot 10mW + 0.3 \cdot 20mW + 0.4ms \cdot 15mW + 0.1ms \cdot 40mW + 19.3\mu J + 9.6\mu J + 0.3ms \cdot 20mW = 53.9\mu J$. This is an improvement of 24.1%, while the first schedule achieved only a reduction of 7.6%. Note, although the schedule in Figure 4.4(a) shows less slack ($0.3ms$) than the one in Figure 4.3(a) ($0.4ms$), its energy reduction is significantly higher with 16.5%. Thus, more slack does not necessarily lead to increased energy savings. This is due to the particular power dissipations when executing the different tasks. In summary, this example has demonstrated how important it is to take into consideration the power variations during the schedule optimisation. In the presence of power variations it is not always true that an increased slack time results in lower energy dissipation, hence an energy-conscious scheduling technique should consider the power variation model.

4.1.2   Background

This section provides brief background information and introduces the terminology that is used throughout the rest of this chapter. This regards scheduling techniques (Section 4.1.2.1) as well as genetic algorithms (Section 4.1.2.2).

4.1.2.1   Scheduling Techniques

In general, scheduling techniques can be broadly classified into two categories: on-line (dynamic) scheduling techniques [44, 125] and off-line (static) scheduling techniques [16, 91, 111, 125]. The technique in the first class dynamically re-calculates the priorities of tasks during run-time of the application, i.e., the schedule can be changed during execution. Obviously, such approaches consume power and time during execution; time which could otherwise be utilised by DVS to lower the energy dissipation. In the second class, a static schedule is calculated once before the application is executed (pre-run-time), i.e., the execution order of tasks and communications is maintained unchanged during run-time, hence, the power and time overhead is avoided. On-line scheduling are advantageous when no or only little knowledge is given about the tasks that have to be performed. In this case, the schedule can be dynamically adapted to the temporal needs of the application. However, embedded systems are often application specific, i.e., they execute a fixed set of tasks that is *a priori* known. Therefore, an adaption during run-time is not crucial for most embedded systems. In addition, off-line scheduling can guarantee that tasks deadlines are met during run-time, while this is not generally the case for dynamic scheduling [125]. Due to these reasons, the techniques introduced in this book concentrate on static scheduling.

Static scheduling for distributed systems that execute tasks with interdependencies has been intensively studied [16, 30, 91, 111, 121, 138, 158]. This scheduling problem belongs to the class of NP-hard problems [62], i.e., the search space for scheduling problems of realistic size is huge ($N!$, where N

is the number of tasks and communication). Hence, finding an optimal solution is extremely computational expensive. For this reason most scheduling techniques rely on heuristic methods that produced not necessarily optimal solutions, but solutions of high quality. One of the most widely used scheduling heuristics is list scheduling (LS). List scheduling algorithms take scheduling decisions based on task priorities. They maintain one or more ready-lists which contain tasks that are ready to be scheduled. A static schedule is constructed by scheduling the ready task with the highest priority as soon as the eligible PE becomes available. Thereby, the assignment of priorities defines the task execution order. Most traditional LS approaches use various sophisticated algorithms to calculate these task priorities either statically [30, 158] (before list scheduling) or dynamically [90, 111, 138] (re-calculation after each scheduling step). Another method for the determination of priorities is used in genetic list scheduling algorithms (GLSA) [46, 66]. In contrast to list scheduling techniques, which produce a single schedule, GLSAs construct and evaluate many different schedules during an iterative optimisation process. However, most of this research on LS and GLSA concentrates on performances aspects only, i.e., the system schedule is solely optimised to achieve the highest possible throughput or to meet the imposed task deadlines. On the other hand, the genetic list scheduling approach presented in this chapter (Section 4.1.3) aims to find schedules that simultaneously meet the system performance requirements and reduce the energy consumption by effectively exploiting the DVS-PEs. As the name suggests, genetic list scheduling approaches are based on genetic algorithms, hence, the following section briefly introduces the terminology and functionality of genetic algorithms.

4.1.2.2 Genetic Algorithms

Genetic algorithms (GAs) have been the subject of numerous investigations in the last decades, and they have been proven to solve different search and optimisation problems successfully [18, 64]. By imitating and applying the principles of natural selection and "survival of the fittest" on a population pool (several solution candidates), they are able to evolve (optimise) solutions to real-world problems. Each solution (individual) of the problem to be solved is encoded as a string (chromosome) and is associated with a solution quality (fitness). Based on their fitness, the individuals are ranked within the solution pool. In each iteration (generation) the algorithm selects upon the highest ranked individuals and gives them the opportunity to reproduce by mating (crossover) with different individuals of the population. This results in new individuals (offsprings) inheriting certain properties and features of the parent individuals, thus potentially increasing the probability to have higher solution quality. The produced offsprings replace the least ranked solutions in the population, which die out. New individuals are not only generated by means of crossover, but also by randomly changing (mutation) values (genes) of a chromosome, occasionally. This provides an additional opportunity to enter unexplored regions of the search space. The GA iterates until a certain stop criterion is fulfilled, for example, the maximum search time has been exceeded or no further improvement can be made.

4.1.3   Genetic List Scheduling Algorithm

In this section a new scheduling technique for energy minimisation through DVS is introduced, which addresses the problem motivated in Section 4.1.1, i.e., scheduling under the consideration of the power variation model. As mentioned in the beginning of this chapter, the presented co-synthesis approaches splits task mapping and scheduling into two separate optimisation steps, although some synthesis techniques combine these steps within a single optimisation [30, 46, 66, 83]. The decision to separate task mapping and scheduling into two "independent" optimisations is based on the following two reasons:

(a) The combination of list scheduling and mapping algorithms decide upon task priorities which task is to be scheduled next, but at this point it is *not* known where to execute the chosen task. Therefore, the execution time and power dissipation of the task are unknown as well. In this context, it is the duty of the scheduler to make a "greedy" mapping decision based on the power and time values with respect to the design objectives. However, DVS influences the execution times and power dissipations, hence, the mapping decision made upon the static values might proof to be wrong, especially from the energy reduction point of view. Separating the scheduling and mapping into two iterative optimisations overcomes this problem since the mapping is given before a schedule is constructed.

(b) Due to the constructive nature of list scheduling and mapping algorithms a solution is constructed one by one. This results in a greedy approach, which is likely to get trapped at low quality or infeasible solutions in the presence of tight area and timing constraints. For instance, it is likely that the scheduler maps early tasks to fast hardware since hardware area is available. Nevertheless, the scheduling algorithm cannot look ahead and hence no area might be left when later, timing critical tasks arrive. A solution to overcome this problem was presented in [83]. However, this approach neglects issues related to power and a straight-forward enhancement towards DVS is not possible due to multiple competing design goals. Nevertheless, by splitting the problem into two steps, this greediness problem is avoided and the advantage of an increased search space, which is explored iteratively, can be leveraged. Clearly, increasing the search space results in higher optimisation times, however, it will be shown in Section 4.1.4 that these times are still reasonable.

The presented scheduling algorithm for the DVS problem under the consideration of the power variation model uses a genetic list scheduling approach to optimise the execution order of the tasks towards energy reduction and timing feasibility. It has been shown that the combination of genetic and list scheduling algorithms provides a powerful tool for the synthesis of multiprocessor systems. However, the proposed implementation varies in two fundamental issues from previous research [46, 66]:

- Instead of optimising the schedule solely for performance (reducing the make-span), the proposed framework considers additionally the issue of energy minimisation with respect to DVS under the power variation model.

- The algorithms described in [46] and [66] employ a list scheduler which determines not only the execution order of tasks, but also their mapping. This combination is avoided in order to limit the greediness of the algorithm which would affect the solution quality.

As opposed to constructive list scheduling technique, genetic list scheduling approaches do not determine *one schedule* using a sophisticated algorithm for the priority assignment, but they construct and evaluate *many different schedules* during an iterative priority optimisation process. By encoding the task priorities into a priority string, it becomes possible to utilise genetic operators, such as crossover and mutation, to change task priorities and hence generate new scheduling solutions using static list scheduling. The principles behind genetic list scheduling approaches are outlined next.

Principle of Genetic List Scheduling

Genetic list scheduling approaches combine fast constructive list scheduling techniques with the optimisation power of genetic algorithms. The basic idea behind list scheduling is shown in Figure 4.5, which outlines the construction of a schedule for a single processor system. Consider the task graph with annotated priorities in Figure 4.5(a). In the initial scheduling step all tasks with no ingoing edges are placed into a ready list[1], as shown in Figure 4.5(b), Step 1. For the given task graph this is solely task τ_0. Being the only task in the ready list, task τ_0 is scheduled. After finished its execution, the tasks τ_1, τ_2, and τ_3 become eligible for scheduling (due to their data dependency on τ_0); hence, they are placed into the ready list in decreasing order of their priorities (Scheduling Step 2). At this point τ_3 represents the task with the highest priority (9), hence it is scheduled in Step 2. Having scheduled task τ_3, task τ_5 becomes ready and thus it is placed into the ready list. These scheduling procedure is repeated until no tasks are left in the ready list. Since each scheduling step schedules one task, seven iterations are necessary. The final schedule is shown in Figure 4.5(c). Clearly, different assignments of priorities result in different schedules. This is where the genetic algorithm comes into the play.

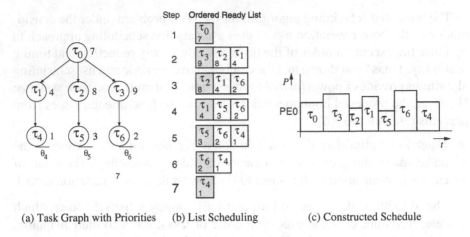

(a) Task Graph with Priorities   (b) List Scheduling   (c) Constructed Schedule

Figure 4.5.   List scheduling

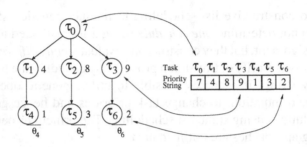

Figure 4.6.   Task priority encoding into a priority string

By encoding the task priorities into a priority string, as shown in Figure 4.6, it becomes possible to apply an genetic algorithm-based optimisation. The genetic algorithm aims to find an assignment of priorities that leads to a schedule solution of high quality in terms of timing behaviour and exploitable slack time. The main principles behind the genetic list scheduling algorithm are illustrated in Figure 4.7. These principles are based on two strategies: crossover and mutation.

Crossover Example

Out of an initial population pool that contains six priority candidate strings, the strings 1 and 3 are selected. Offsprings are produced by replacing parts of the first parent string with parts of the second parent sting. Hence, crossover results in two new offsprings (child 1 and child 2). These new priorities are used to schedule the activities, in order to determine their quality. According to the quality, the produced strings are inserted into the solution pool. By selecting high quality strings for crossover, the chances to evolve priority strings of even

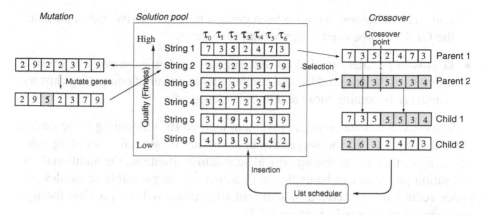

Figure 4.7.   Principle behind the genetic list scheduling algorithm

higher quality are increased. The mating of two strings is carried out with
respect to an arbitrarily selected crossover point.   □

Mutation Example
In order to enter into unexplored regions of the scheduling space, the genetic
algorithm mutates individuals of the solution pool occasionally (with a low
probability). The mutation is carried out by randomly changing genes of a
randomly selected string. For instance, in Figure 4.7 string 2 is selected and its
third gene is manipulated. The modified string is then reinsert into the solution
pool.   □

Both crossover and mutation are applied during the iterative execution of the
genetic algorithm. The algorithm terminates after a stop criterion is fulfilled.

Genetic list scheduling approaches (GLSA) have several advantages compared to traditional (constructive) list scheduling methods. These advantages
can be summarised as follows:

- Multi-objective optimisation is an important feature that is supported by
  genetic algorithms. It provides the opportunity to simultaneously optimise
  the implementation towards competing goals and so allows the system de-
  signer/architect to choose between several suitable implementations with
  different properties.

- The enlarged search space (at most $(|\mathcal{T}| + |\mathcal{C}|)!$ different schedules can be
  produced) provides the opportunity to find solutions of potentially higher
  quality.

- There is a large freedom to trade off between acceptable synthesis time
  and solution quality; as opposed to constructive techniques, where only one
  solution is produced rapidly. In addition, the search process can be initialised

with such solutions that have been created by constructive techniques, and the GLSA can be used to further improve these.

■ GAs with parallel populations and migration schemes provide a powerful approach to leverage additional computational power of computing clusters, which are becoming more and more widespread.

A drawback of any iterative improvement method for scheduling is the timing overhead involved in the successive construction of several scheduling solutions. Nevertheless, in timing critical scheduling situations the additional optimisation potential can be exploited to achieve timing feasible schedules and hence reduce the embedded system cost (this claim will be justified through extensive experiments in Section 4.2.3).

Implementation Details

The following introduces a DVS optimised genetic list scheduling algorithm (EE-GLSA [2]) that addresses the execution order of tasks and communications under the consideration of the power variation model. The pseudo code of the **EE-GLSA** is show in Figure 4.8, which describes its features and implementation details. The solution pool (25 individuals) of the first generation is initialised (step 01) half and half by mobility-based [158] and randomly generated priorities (with values between the lowest and highest mobility), respectively. The mobility of a task is given by the difference between the as-late-as-possible (ALAP) start time and the as-soon-as-possible (start time) of a task [158]. This initial population was empirically found to be a good starting point, leading to fast convergence (i.e., low optimisation times). The algorithm then enters the main schedule optimisation loop (step 02–10), which is repeated until no improvement of at least 1% (with respect to the best found feasible schedule) is made within 10 generations. Each iteration of the loop goes successively through the following steps. All new priority candidate strings in the solution pool are used by the list scheduling algorithm to generate schedules at nominal supply voltage (step 02). The implemented list scheduler relies solely on the task priorities to make schedule decisions, i.e., no other techniques, like e.g. hole filling, are used to optimise the schedule. Although such techniques can improve the timing behaviour by eliminating idle periods in the schedule, the used list scheduler dissociate from them since the DVS technique exploits exactly these idle times. For a simple example consider the task set given in Figure 4.9, mapped onto an architecture build out of two DVS-PEs. The task priorities are given on the right side of each task. According to these priorities, a static list scheduler can generate a feasible schedule, as shown in Figure 4.9. It can be observed that the tasks τ_0, τ_1, and τ_2 can be scaled only a small amount until deadline d_2 is met. However, task τ_4 can utilise the idleness before task τ_2 starts execution, and task τ_3 can be scaled until the deadline $d_{3,4}$ is met. Let

Algorithm: *EE-GLSA*

Input: - task graph TG
          - mapping and execution properties corresponding to
           the mapping
Output: - timing and energy optimised schedule

01: **Initialisation:** Create initial population pool P of priority strings, half randomly generated and half based on mobility.
02: **Perform List Scheduling:** Generates, for each member of the solution pool, a schedule based on the corresponding priority string.
   a) Assign task priorities from the property string
   b) Invoke list scheduler without hole filling
03: **Perform Voltage Scaling:** Invoke the generalised DVS technique, calculating supply voltages for each task executed on a DVS-PE. This is done under the consideration of the individual power dissipation of tasks.
04: **Assign Fitness:** Compute fitness of each individual in the population pool.
   a) Calculate timing penalty
   b) Calculate energy based on the supply voltages
   c) Derive fitness based on energy and timing penalty
05: **Termination:** If no improved individual (improvement $> 1\%$) has been produces for 10 generations, then terminate. Otherwise, continue with step 06.
06: **Ranking:** Individuals are ranked according to their fitness.
07: **Selection:** According to the size of the generational overlap select individuals for mating. High ranked individuals have a high probability to be selected.
08: **Mating:** Produce two-point crossover between a pair of selected individuals.
09: **Mutation:** Randomly change genes of individuals using a dynamic mutation probability scheme, with exponential decreasing probability during run-time.
10: **Offspring insertion:** Exchange low ranked individuals by newly produced individuals with respect to the size of the generational overlap. Continue with step 02.

Figure 4.8. Proposed EE-GLSA approach for energy-efficient schedules

us consider now the employment of a hole filling technique. In this case the list scheduler would try in its last scheduling step to place task τ_3 into the idle pe-

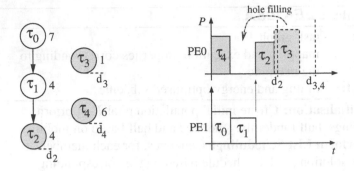

Figure 4.9. Hole filling problem

riod between task τ_4 and task τ_2. This decision, however, is fatally wrong from an energy reduction point of view, since the only available slack for all tasks would be the time between the execution of task τ_2 and the deadline d_2, leaving not much headroom for voltage scaling. On the other hand, if the deadline $d_{3,4}$ would be identical with deadline d_2, then the schedule displayed in Figure 4.9 would become infeasible and the schedule produced with hole filling would represent a better choice. To avoid such a dilemma, the used list scheduler solely makes scheduling decisions based on the task priorities, which are iteratively optimised. Therefore, the proposed **EE-GLSA** is capable of producing both schedule variants discussed above, and it evaluates their suitability with respect to timing and energy aspects to make an appropriate decision.

After the list scheduling has constructed a schedule (step 02), the algorithm proceeds by passing the built schedules to **PV-DVS** algorithm of Chapter 3, which identifies scaling voltages that minimise the energy dissipation (step 03). Note that schedules which exceed hard deadline constraints are still scaled as much as possible and are not excluded from the optimisation. This is done since good solutions are likely to be found as result of transformations performed on invalid configurations. However, a time violation penalty is applied in such cases, as explained next. The scaled schedule is evaluated in terms of deadline violations and energy dissipation including the DVS reductions. Based on this evaluation, the fitness F_S of each schedule candidate is calculated (step 04) using the following equation:

$$F_S = \underbrace{\left(\sum_{\epsilon \in \mathcal{A}} E(\epsilon) \right)}_{\text{Energy diss.}} \cdot \underbrace{\left(1 + \frac{\sum\limits_{\tau \in \mathcal{T}_d} DV_\tau^2}{T_{HP}^2} \right)}_{\text{Time penalty}} \tag{4.1}$$

where $\mathcal{A} = \mathcal{T} \cup \mathcal{C}$ defines the set of all activities (tasks as well as communications) and $\mathcal{T}_d$ represents the set of all hard deadline tasks. The first part of the equation is used to calculate the total dynamic energy dissipation of all activities $\epsilon \in \mathcal{A}$. Based upon the type of activity, the energy dissipation can be calculated in the following way,

$$E(\epsilon) = \begin{cases} P_{max}(\epsilon) \cdot t_{nom}(\epsilon) \cdot \frac{V_{dd}^2(\epsilon)}{V_{max}^2(\epsilon)} & \text{if } \epsilon \in \mathcal{T}_{\text{DVS}} \\ P_{max}(\epsilon) \cdot t_{nom}(\epsilon) & \text{if } \epsilon \in \mathcal{T} \setminus \mathcal{T}_{\text{DVS}} \\ P_C(\epsilon) \cdot t_C(\epsilon) & \text{if } \epsilon \in \mathcal{C} \end{cases} \qquad (4.2)$$

where P_{max} and t_{nom} refer to the power dissipation and execution time of tasks at nominal supply voltage, respectively. V_{dd} is the scaled supply voltage, $\mathcal{T}_{\text{DVS}}$ represents the set of all tasks mapped to DVS-PEs, and P_C and t_C denote the power and execution time of communication activities. It should be noted that the power dissipations and the execution times of the tasks depend on the found scaling voltages V_{dd}, which have been calculated using the **PV-DVS** algorithm introduced in Chapter 3. The second part of the fitness function introduces a penalty factor due to deadline violations of deadline tasks,

$$DV_\tau = \max \left(0, (t_S(\tau) + t_{exe}(\tau)) - t_d(\tau) \right) \qquad (4.3)$$

where $t_S(\tau)$ and $t_{exe}(\tau)$ denote the start time and the execution time (possibly scaled) of task τ, and $t_d(\tau)$ refers to the task deadline. T_{HP} represents the period of the hyper task graph (least common multiplier of all task graph periods), used to relate the deadline violation. Squaring has been applied in order to apply a higher penalty to larger violations of imposed deadlines.

By guiding the optimisation with this equation, the search for schedules is pushed into regions where low energy *and* feasible schedules are likely to be found. The algorithm then checks the halting criterion, as mentioned above. If the end of the optimisation has not been reached, the algorithm continues (step 05) and the new priority candidates are ranked (step 06) and inserted into the solution pool based on their fitness values. Low ranked individuals of the pool are replaced (step 07) by new ones, which are generated through genetic crossover (step 08) and mutation (step 09). The employed GA is of steady state type, that is, not all of the individuals in the solution pool are replaced with each iteration (step 10). Steady state GAs where used due to their perfor-mance advantages compared to generational GAs, as indicated in [128]. The generational gap was set to 50%, i.e., half of the individuals in the solution pool survive unchanged in each generation. The crossover is carried out by means of a random two-point crossover [64]. To avoid a premature convergence towards suboptimal schedules, the idea of a dynamic mutation probability is leveraged [57]. This approach gives the algorithm the additional feature and capability to easily escape local minima in the beginning of the optimisation run. The mu-tation probability follows the equation $1/\exp(N_S \cdot 0.05)$ and is never allowed to drop below 15%. N_S denotes the current generation during the schedule op-timisation. The generated offsprings are inserted into the population, resulting in a new generation. At this point, the next iteration is invoked and so different schedules are tried out.

4.1.4   Experimental Results: Schedule Optimisation

This section demonstrates through several benchmark experiments that the genetic list scheduling algorithm EE-GLSA introduced in this chapter achieves high energy savings, particularly when considering power variations during the optimisation. The EE-GLSA has been implemented on a PentiumIII/750MHz PC using a publicly available library of genetic algorithms [150]. All reported results represent average values that have been obtained over ten optimisation runs. In addition to the benchmarks (a–c) described in Chapter 3, the experiments conducted in this section also concern the benchmarks set used in [68].

(d) Gruian's and Kuchcinski's task graphs [68] represent two sets (TG1 and TG2) of 30 randomly generated, communicating tasks with tight deadlines (determined by a critical path scheduling algorithm). These graphs show a high degree of parallelism and are mapped to architectures built of 3 or 10 identical DVS-PEs, assuming constant power consumption. These PEs are multi-voltage processor able to run at $3.3V$, $2.5V$, $1.7V$, and $0.9V$, while the threshold voltage V_t is $0.4V$.

The following experiments are split into two sections. The first section concentrates on experiments that highlight the importance of considering power variations during scheduling in order to increase the energy-efficiency. While the second section compares the genetic algorithm based iterative schedule optimisation with constructive list scheduling techniques.

Power Variation Experiments

As demonstrated in the motivational scheduling example at the beginning of this section, power variations can have a significant influence on the suitability of a schedule when DVS is applied. This section assesses this influence through the usage of several benchmarks experiments. For this purpose Table 4.2 compares two different approaches; both are based on the same genetic list scheduling technique, however, the first approach considers the fixed power model, while the second employs the power variation model. The table shows for both approaches the achieved energy reductions and computational overheads when applied to the benchmark set tgff. The achieved energy reductions using the fixed power model varied between 1.35% in the case of tgff6 and 51.89% in the case of tgff3. However, in all cases the energy could be further reduced when considering the power variations using PV-DVS. In these cases the reductions varied between 1.60% (tgff6) and 69.21% (tgff3). Accordingly, up to 17.32% higher savings could be achieved solely by reordering the execution of tasks and communications.

	GLSA + EVEN-DVS (fixed power model)			EE-GLSA (GLSA + PV-DVS) (proposed power variation model)		
Example	*Energy (μJ)*	*CPU time (s)*	*Total Reduction (%)*	*Energy (μJ)*	*CPU time (s)*	*Total Reduction (%)*
Tgff1	190.74	0.12	46.27	102.37	0.14	71.16
Tgff2	572920.30	0.20	22.91	545450.95	0.27	26.61
Tgff3	266906.88	0.33	51.89	170837.88	3.11	69.21
Tgff4	377444.70	0.24	12.55	375778.35	0.97	12.94
Tgff4_t	405473.20	0.25	6.06	396579.23	0.62	8.12
Tgff4_fixed	127867.31	0.26	27.65	124419.18	1.00	29.60
Tgff5	3721136.80	0.37	11.13	3450291.60	2.41	17.60
Tgff6	1399967.70	0.23	1.35	1396445.00	0.25	1.60
Tgff7	1924999.80	0.20	24.47	1797520.40	0.27	29.47
Tgff8	1722056.40	0.19	10.01	1648321.50	0.20	13.86
Tgff9	829607.95	0.18	16.76	774993.70	0.26	22.24
Tgff10	45324.61	0.17	34.65	44529.05	0.22	35.79
Tgff11	3755206.40	0.22	13.67	3621739.60	0.42	16.73
Tgff12	2212405.00	0.34	4.49	2198978.20	3.73	5.07
Tgff13	2342891.80	0.28	19.56	2315765.60	0.80	20.49
Tgff14	11891.05	0.21	23.44	11752.85	0.25	24.33
Tgff15	61271.38	0.41	2.13	60129.32	1.07	3.96
Tgff16	2492364.60	0.26	28.68	2449747.20	0.55	29.90
Tgff17	18922.85	0.27	19.34	18249.41	0.56	22.21
Tgff18	1724421.20	0.14	6.87	1421224.40	0.16	23.25
Tgff19	4514.75	0.17	23.98	4357.35	0.16	26.63
Tgff20	42703.58	0.19	45.02	37222.68	0.60	52.08
Tgff21	2983043.80	0.56	6.13	2578046.00	3.92	18.87
Tgff22	4664876.00	1.19	19.87	4119749.20	3.44	29.23
Tgff23	9826644.00	0.64	15.05	8855575.20	21.25	23.44
Tgff24	5240976.80	0.59	2.08	4881187.60	10.48	8.80
Tgff25	4922084.80	0.39	14.18	4545249.60	1.91	20.75

Table 4.2.   Experimental results obtained using the *fixed power model* and the *power variation model* during voltage selection; both integrated into a genetic list scheduling algorithm

Clearly, an advantage of the EVEN-DVS technique (fixed power model) is its linear time complexity, when compared to the quadratic complexity of the PV-DVS algorithm (power variation model). This is also reflected in the reported optimisation time of Table 4.2. As it can be observed, the optimisation times for the approach using a fixed power model varied from $0.12s$ in the case of tgff1 to $1.19s$ in the case of tgff22. The optimisation times for the same benchmarks optimised under the power variation model vary between $0.14s$ (tgff1) and $21.25s$ (tgff22). It is a classical trade-off between optimisation time and solution quality.

Comparison with Constructive List Scheduling Technique

In order to assess the quality of the new EE-GLSA technique to achieve high energy savings not only due to the consideration of the power variation model, this section provides a comparison with constructive list scheduling techniques. The first comparison of the EE-GLSA is against a mobility-based (i.e., performance-driven) constructive list scheduling technique [158]. The results presented in Table 3.4 (page 52) have been generated using a mobility-based list scheduling, while the results Table 4.2 are based on the genetic list scheduling approach. Hence, comparing the results of both tables is equivalent to a comparison between both scheduling approaches. Take, for instance, example tgff2. When considering a constructive list scheduling based on mobility and a voltage scaling based on PV-DVS, an energy reduction of 7.97% can be achieved (Column 7 in Table 3.4). Nevertheless, applying the genetic list scheduling approach combined with PV-DVS, the energy is reduced by 26.61% (Column 7 in Table 4.2). Note, these savings are solely achieved through an improved execution order of tasks and communications, since in both cases the same voltage scaling technique (PV-DVS) has been applied. Overall it can be observed that the genetic list scheduling technique yields higher energy savings in most of the examples. In fact, only in the case of benchmark tgff6 no further improvement could be achieved through genetic list scheduling. This is due to the timing critical execution of tasks, which offers no energy reduction potential. Certainly, the GA based schedule optimisation introduces a computational overhead, which results in a necessary trade-off between solution quality in terms of energy dissipation and synthesis time. Nevertheless, the EE-GLSA can produce solution of at least the same quality as the mobility-based approaches in the "same" time, by placing a string with priorities based on the mobility criterion into the initial population of the EE-GLSA.

To further confirm the quality of the EE-GLSA, it is next compared to the DVS scheduling technique proposed by Gruian et al. [68]. This comparison is carried out using the benchmark collections TG1 and TG2, which contain 60 task graph examples with tight deadlines. The individual graphs capture highly parallel tasks with relatively few communications. Nevertheless, for experimental purpose these benchmarks represent a valuable choice. The reported energy reductions in [68] for these benchmarks are 28% and 13%. Tables 4.3 and 4.4 present the results obtained using the genetic list scheduling algorithm EE-GLSA, which is driven by the PV-DVS algorithm.

The table separates both benchmark collections. Although the examples do not allow the EE-GLSA approach to leverage power variations, since the specified power values are constant, the achieved average energy reductions for TG1 and TG2 are 41.16% and 18.82% (Column 5 and 11, last row), respectively. This is an improvement of 13.16% and 5.82%, which indicates the effectiveness of the proposed optimisation technique, even when using constant power benchmark examples. However, since the results in [68] are obtained using multi-voltage PEs rather than variable-voltage PEs, additional experiments have been conducted, using the same discrete PE voltages ($V_{dd} =$

Example	No. of nodes & edges	Contin. Energy Dissip.	CPU time (s)	Contin. Reduc. (%)	Discrete Reduc. (%)
r000	30/24	474783	6.09	40.56	37.35
r001	30/19	395964	10.97	47.87	44.11
r002	30/23	523822	2.97	29.67	26.71
r003	30/20	504252	12.67	49.31	45.83
r004	30/18	541220	4.32	38.98	35.13
r005	30/17	432787	8.27	41.89	39.10
r006	30/16	592832	4.33	34.25	31.03
r007	30/23	551119	6.30	34.23	31.07
r008	30/20	590249	4.20	31.56	28.69
r009	30/14	369454	6.98	45.76	41.71
r010	30/25	315060	13.43	55.35	51.48
r011	30/23	421029	11.36	47.61	43.83
r012	30/20	675304	5.51	32.11	29.06
r013	30/20	543053	5.87	33.64	29.87
r014	30/19	429695	12.10	50.73	46.36
r015	30/17	420428	11.15	48.31	44.53
r016	30/24	481347	5.99	35.37	32.21
r017	30/27	510259	13.37	42.47	38.97
r018	30/17	426713	9.70	39.10	36.08
r019	30/23	577211	7.18	33.82	29.61
r020	30/24	395536	10.41	46.89	42.66
r021	30/15	675748	5.49	31.39	28.64
r022	30/24	457843	13.59	44.05	40.63
r023	30/20	379382	11.78	52.50	47.87
r024	30/16	530941	6.23	35.88	31.70
r025	30/15	301150	15.67	60.60	57.74
r026	30/16	640655	3.27	26.96	23.06
r027	30/16	483131	2.49	32.47	29.10
r028	30/20	430367	12.06	47.09	43.43
r029	30/26	365148	11.46	44.39	40.65
Average Values:			**8.51**	**41.16**	**37.61**

Table 4.3. Experimental results obtained using the generalised, DVS optimised scheduling approach for benchmark example TG1

$\{0.9V, 1.7V, 2.5V, 3.3V\}$, while $V_t = 0.4V$). As outlined in Section 3.2.2, each continously selected voltage (using the PV-DVS algorithm) can be split into its two neighbouring discrete voltages considering the available voltages of the multi-voltage PE. The corresponding run-times at each voltage are calculated using the Equations (3.5) and (3.6). The results of the discrete voltage optimisation are shown in Tables 4.3 and 4.4, see columns with the headings "Discrete Reduc.". For the two benchmark sets the achieved average energy reductions are

Example	No. of nodes & edges	Contin. Energy Dissip.	CPU time (s)	Contin. Reduc. (%)	Discrete Reduc. (%)
r000	30/22	659972	0.41	20.30	17.24
r001	30/23	809191	0.55	13.99	11.39
r002	30/14	522578	1.76	25.94	24.35
r003	30/22	604127	1.06	20.97	18.11
r004	30/23	523668	1.18	25.27	22.23
r005	30/17	694278	0.94	20.40	17.13
r006	30/16	632867	1.64	20.27	17.66
r007	30/23	717256	0.42	7.94	5.34
r008	30/18	513749	1.70	34.06	30.32
r009	30/18	763976	0.81	14.80	10.92
r010	30/18	669359	0.91	21.04	17.66
r011	30/21	644326	0.90	20.79	16.58
r012	30/18	660552	0.29	13.03	9.80
r013	30/20	697671	0.34	10.45	7.64
r014	30/22	658146	0.35	17.09	14.12
r015	30/15	738860	0.85	13.34	9.85
r016	30/20	538039	1.32	22.12	18.98
r017	30/16	654964	1.53	22.29	19.42
r018	30/20	706124	0.90	19.04	16.91
r019	30/19	693263	1.56	19.15	15.80
r020	30/23	684600	0.58	16.34	13.39
r021	30/20	667156	0.29	13.28	10.44
r022	30/22	454777	1.15	33.71	29.94
r023	30/18	579582	0.96	18.43	15.59
r024	30/19	633231	0.86	22.15	18.89
r025	30/21	859903	0.37	15.63	12.74
r026	30/18	621194	0.60	18.73	15.60
r027	30/18	722126	0.56	11.22	8.00
r028	30/15	555117	1.64	26.91	24.95
r029	30/18	769308	0.55	5.99	3.93
Average Values:			**0.90**	**18.82**	**15.83**

Table 4.4. Experimental results obtained using the generalised, DVS optimised scheduling approach for benchmark example TG2

37.61% and 15.83%, respectively, which represent improvements of 9.61% and 2.83%. Note that these reductions were obtained on benchmarks which do *not* show any power variations and so this optimisation feature of the proposed DVS algorithm stays unexploited. The achieved improvements are due to the fact that the proposed iterative GA based scheduling approach is able to explore a large space of potentially low-energy schedules, as opposed to the constructive list scheduling approach used in [68]. Regarding the computational times, Gruian *et al.* reported average times for the 30-node task graphs of $10s$ to $120s$ (on $440MHz$, UltraSparc IIi, 256MB Workstation), while the proposed algorithm executes on average in $0.9s$ to $8.51s$ (on $750MHz$, Pentium III PC, 128MB). These longer execution times of the constructive technique can be explained by the fact that the scheduling used in [68] has to re-schedule the tasks when no feasible schedule is found, i.e., the task priorities are re-adjusted. Further, the UltraSparch Workstation provides approximately 2–3 less computational power compared to the PentiumIII PC according to SPECint performance evaluation [7]. However, the optimisation times indicate an advantage of the presented EE-GLSA technique.

In summary, this section has shown that significant improvements in terms of energy savings can be made by optimising the execution order of tasks and communications. In particular when compared to constructive techniques, the new EE-GLSA can achieve higher energy savings due to the effective exploration of the scheduling search space.

4.2   Optimisation of Task and Communication Mapping

Section 4.1 has shown that substantial energy savings (up to 17.32% were observed) can be achieved through an schedule optimisation that improves the execution order of tasks and communications not only towards performance goals, but additionally towards the effective utilisation of dynamic voltage scaling. Clearly, application mapping (Section 1.3.2) and activity scheduling (Section 1.3.3) are two heavily interrelated co-synthesis steps. For instance, mapping parallel tasks onto a single processing element would necessitate to execute these tasks one after the other (sequentially). On the other hand, mapping these tasks to different processing elements would allow to execute the tasks in parallel. This means that the mapping of tasks and communications has an important influence on the schedulability as well as on the utilisation of DVS and hence should be subject to optimisation from a timing and energy point of view. This section introduces a novel *two-step* approach that aims to improve the mapping towards these goals. Conceptually, the mapping approach *separates* the optimisation of task and communication mapping, i.e., both assignments are carried out in isolation of each other, as illustrated in Figure 4.1 (page 62). Correspondingly, this section is divided into two sections. Section 4.2.1 introduces a task mapping based on genetic algorithms that has been adapted to suit the particular problem of optimising the design for the effective exploitation of the DVS-PEs. Section 4.2.2 proposes a new method that extends the scheduling technique outlined in Section 4.1 to a combined optimisation of communication mapping and scheduling. Again, this technique aims at performance improvement as well as DVS utilisation. Using these techniques, the influence of mapping on the achievable energy savings through DVS is analysed in Section 4.2.3 for a set of benchmark experiments.

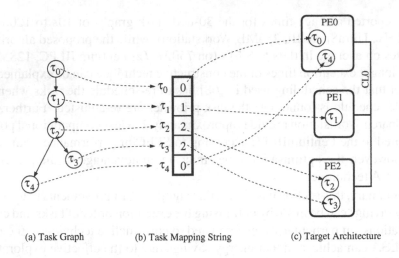

(a) Task Graph   (b) Task Mapping String   (c) Target Architecture

Figure 4.10.   Task mapping string describing the mapping of five tasks to an architecture

4.2.1   Genetic Task Mapping Algorithm

The task mapping step determines which PE carries out which task. Thereby, it determines the execution time and power dissipation of a task at nominal supply voltage and further the area requirement in terms of bytes or gates, whether a task is implemented as software or hardware. The goal of the mapping optimisation step is to distribute the tasks among the processing elements that form the distributed architecture, including the DVS-enabled PEs, such that the energy dissipation is minimised and feasible designs in terms of timing behaviour and area constraints are achieved. As mentioned in Section 2.4, task mapping has been intensively researched over the last decade. And similar to the scheduling problem, it belongs to the class of NP-hard problem [62]. That is, optimal solution for realistic problem sizes may only be found through extremely computational expensive processes. One effective way to address this problem is the usage of genetic algorithms for task mapping [47, 49]. Nevertheless, previous approaches did not consider the presence of DVS-PEs. In addition, as opposed to the mapping approach introduced in [47, 49], where solely the mapping of tasks is optimised by a GA, the approach presented in this book uses two independent GAs to find improved solutions for the mapping of communications and tasks. This section focuses on task mapping.

In GA-based task mapping approaches, solution candidates (potential mappings) are encoded into mapping strings [49], as shown in Figure 4.10. Each gene in these strings describes a mapping of a task to a processing element. For instance, task τ_4 in Figure 4.10 is mapped to PE0. Similarly to the genetic algorithm used for the schedule optimisation (Section 4.1.1), the genetic task mapping algorithm (**EE-GTMA**) evolves a population of possible mapping so-

Algorithm: *EE-GTMA*

Input: - task graph TG
            - technology library (execution times, power dissipations)
            - allocated arcitecture
 Output: - timing, area, and energy optimised mapping

01: **Initialisation:** Create initial population pool P_m of mapping strings, generated randomly   02: **Perform Mapping:** Generates, for each member of the solution pool, a mapping based on the corresponding mapping string. Specifices the task properties such as execution time, power dissipation, etc.   03: **Invoke EE-GLSA:** Invoke the schedule optimisation to, determine a suitable and energy efficient schedule for the current task mapping.   04: **Assign Fitness:** Compute fitness of each individual in the population pool.   a) Calculate area penalty   b) Derive fitness based on are penalty and the schedule fitness.   05: **Termination:** If no improved individual (improvement $> 1\%$) has been produces for 10 generations, then terminate. Otherwise, continue.   06: **Ranking:** Individuals are ranked according to their fitness.   07: **Selection:** According to the size of the generational overlap select individuals for mating. High ranked individuals have a high probability to be selected.   08: **Mating:** Produce two-point crossover between a pair of selected individuals.   09: **Mutation:** Randomly change genes of individuals using a dynamic mutation probability scheme, with exponential decreasing probability during run-time.   10: **Offspring insertion:** Exchange low ranked individuals by newly produced individuals with respect to the size of the generational overlap. Continue with step 02.

Figure 4.11. Proposed EE-GTMA approach for energy-efficient task mappings

lutions towards high quality implementations. The genetic algorithm that is employed to work on the task mapping strings is described in Figure 4.11. As typical in all genetic algorithms, the EE-GTMA applies ranking, selection, crossover, mutation, and offspring insertion in order to evolve an initial solution pool. The key feature of the EE-GTMA, however, is the invocation of the genetic list scheduling algorithm (EE-GLSA) for each mapping candidate, which allows to calculate parts of the fitness function that guides the optimisation. More precisely, the scheduling fitness F_S is used within the task mapping fitness F_M, as shown in the following equation:

$$F_M = F_S \cdot \prod_{\pi \in \mathcal{P}} AP_\pi \qquad (4.4)$$

$$AP_\pi = \begin{cases} 1 & \text{if } AA_\pi \geq SA_\pi \\ k \cdot \left(\frac{SA_\pi}{AA_\pi} - 1 \right) + 1 & \text{otherwise} \end{cases} \qquad (4.5)$$

where F_S is the schedule fitness (Equation (4.1), Section 4.1.3) based on the DVS reduced energy dissipation and a time penalty as outlined in Section 4.1.3. AP_π assigns an area penalty for each PE exceeding its area constraints as given in Equation (4.5). The used area is denoted SA_π and the maximal available area is represented by AA_π (either as memory or silicon area depending on the implementation in SW or HW). If the available area AA_π is not exceeded, it is not necessary to assign an area violation penalty for the particular processing element π, hence, F_S is multiplied by one. On the other hand, if the area constraint is exceeded, the used area SA_π and the available area AA_π are related and multiplied by a constant k, which allows to adjust the aggressiveness of the penalty. Through extensive experiments, a value of 0.02 was found to be a good choice for the constant k, which was sufficiently high to avoid infeasible results at the end of the mapping optimisation. However, this value for k is still low enough to allow infeasible solutions to survive sometimes in order to increase the population diversity and avoid a premature convergence of the GA towards solutions of unnecessary low quality. In this way, it is possible to stimulate the placement of functionality onto the distributed PEs such that energy is minimised, while timing and area constraints are respected. The parameters of the GA for the task mapping were set as follows: The population size was set to 50, the minimal dynamic mutation probability was adjusted to 5%, the generational gap comprises 20%, and the initial population pool was filled with random mappings.

4.2.2   Combined Scheduling and Communication Mapping

The mapping approach in the previous section is used for the assignment of tasks to the processing elements only. However, communication issues have a great impact on the timing behaviour of the application and therefore should be considered carefully during the design space exploration [50, 88, 126], in order to find energy-efficient systems. One important decision that has been taken in this regard was the separation of communication mapping from the task mapping within the synthesis approach, i.e., communication and task mapping

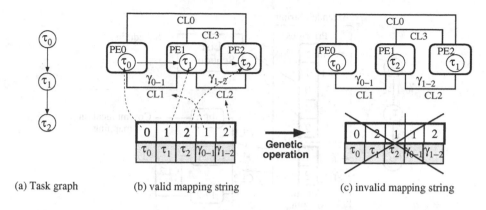

(a) Task graph   (b) valid mapping string   (c) invalid mapping string

Figure 4.12.   Combined optimisation of task and communication mapping

are carried out in two separate optimisation steps (see Figure 4.1). The following example illustrates the reasons behind this decision.

Example

The three tasks and two communications of the task graph shown in Figure 4.12(a) need to be mapped onto a target architecture consisting of three PEs, connected by four CLs. The string shown in Figure 4.12(b) combines tasks mapping (τ_0, τ_1, τ_2) and communication mapping (γ_{0-1}, γ_{1-2}) into one representation. This mapping represents a valid solution since communication γ_{0-1} between the tasks τ_0 and τ_1 is assigned to CL1, which connects the processing elements that accommodate tasks τ_0 and τ_1. In a similar way, the communication γ_{1-2} is mapped onto a communication link (CL0) that interconnets the processing elements of task τ_1 and task τ_2. Let us consider a certain genetic operation which transforms the valid mapping string shown in Figure 4.12(b) to the one in Figure 4.12(c). It can be observed that the mapping of the tasks τ_1 and τ_2 has been modified, while the communication mapping stays unchanged. A quick check upon this mapping indicates that this assignment of activities represents an invalid solution. Consider, for example, the communication γ_{0-1} between task τ_0 and τ_1. Although the tasks are mapped onto PE0 and PE2, which are solely connected through CL0, the communication is mapped to CL1. Hence, this mapping is invalid (if it is considered that only direct communications are allowed, i.e., communications without routing over intermediate PEs). Due to this reason, a combined task and communication mapping approach would produce a *high number of invalid solutions* during the GA based optimisation, which, in turn, would have a negative effect on the convergence of the population towards high quality solutions. To overcome this problem, the proposed mapping approach explicitly separates task and communication mapping. In précis, the introduced communication map-

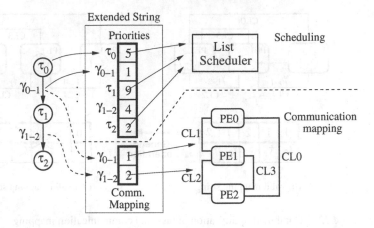

Figure 4.13. A combined priority and communication mapping string

ping technique is carried out in conjunction with the scheduling optimisation within the innermost loop of the proposed synthesis approach (see Figure 4.1). Hence, for each task mapping candidate, the scheduling and communication mapping are simultaneously optimised. In this way it is possible to avoid invalid solutions, since all possible mappings of communication activities onto the communication links are statically known for a particular task mapping. To clarify this consider again Figure 4.13. If the tasks τ_0 and τ_1, for instance, are mapped to PE1 and PE2, respectively, then the communication γ_{0-1} can only be mapped onto CL2 or CL3. Hence, during the optimisation of the communication mapping it is possible to restrict the search to such feasible communication links. The proposed communication mapping, described next, takes advantage of this information to ensure that only valid solutions are produced. Thereby, the search space is restricted to structurally viable solutions only, decreasing significantly the synthesis run-time.

As mentioned above, the presented communication mapping optimisation is carried out in parallel with the schedule optimisation. To explain this strategy consider the extended string representation shown in Figure 4.13, which encodes both a possible schedule and a communication mapping candidate. It can be observed that this string is divided into priority and communication mapping genes. A list scheduler determines an execution order based on the encoded priorities, whilst the mapping of communication activities onto the interconnecting links is given by the communication mapping genes. The combined string representation allows the concurrent optimisation of priorities and communication mappings, using a single genetic algorithm. However, it necessitates a specialised genetic mutation, which operates on the two string parts without interference, i.e., random modifications on priorities need to be considered differently than the modifications on the communication mappings.

On the other hand, standard crossover methods can be directly applied without worrying about feasibility, since the crossover between two strings maintains spatial locality, i.e., priority genes are not mixed with the mapping strings. Nevertheless, during initialisation and mutation of the communication mapping genes only valid values, which result in feasible communication mapping solutions, are allowed. This is not hard to achieve due to the fact that the task mapping precedes the communication mapping (scheduling and communication mapping are carried out in the innermost loop of the synthesis). Thereby, for every communicating pair of tasks the viable CLs are unambiguously specified. Therefore, it is possible to generate random initial chromosomes (random in the sense that a random choice is taken among the possible CLs) that assure proper communication mappings. Similarly, the mutation operator chooses randomly among the valid possibilities only. In order to keep the optimisation time low, the communication mapping string is dynamically adapted to the particular task mapping, as the number of inter-PE communications can changes for each potential task mapping. Thereby, the amount of genes in the mapping string is kept at minimum, avoiding needlessly high synthesis times. Of course, the valid values of each gene change also dynamically in accordance to the task mapping. Note that the presented communication mapping optimisation improves both the timing behaviour as well as the power consumption, since the guiding fitness (Equation (4.1)) accounts for both.

The pseudo code of the combined scheduling and communication mapping algorithm, called **EE-GLSCMA**, is shown in Figure 4.14, an extended version of the **EE-GLSA** algorithm described in Figure 4.8 (page 73). The following modifications can be identified: In Step 01, the proposed combined optimisation of communication mapping and scheduling finds, for all data transfers between tasks, the feasible communication links. In addition to the initial priorities, Step 02 needs to generate random yet feasible communication mappings for the start of the optimisation. During Step 03, which performs the list scheduling, the communication properties are calculated, in order to allow for the consideration of contention over the CLs. Finally, in Step 10, a specialised mutation is carried out, in order to avoid creation of invalid strings.

Algorithm: *EE-GLSCMA*

Input:	- task graph TG
	- task mapping and, correspondingly, the execution properties
Output:	- timing and energy optimised schedule
	- timing and energy optimised communication mapping

01: **Find Feasible Communication Mappings:** Identify possible mappings of communication events to links, depending on the given task mapping.

02: **Initialisation:** Create initial population pool P of combined priority and communication mapping strings. Starting priorities are half randomly generated and half based on mobility, while initial, feasible communication mappings are randomly created.

03: **Perform List Scheduling:** Generates, for each member of the solution pool, a schedule based on the corresponding priority string.
a) Map communication to links
b) Assign task priorities from the property string
c) Invoke list scheduler without hole filling

04: **Perform Voltage Scaling:** Invoke the generalised DVS technique, calculating supply voltages for each task executed on a DVS-PE. This is done under the consideration of the individual power dissipation of tasks.

05: **Assign Fitness:** Compute fitness of each individual in the solution pool.
a) Calculate timing penalty
b) Calculate energy based on the supply voltages
c) Derive fitness based on energy and timing penalty

06: **Termination:** If no improved individual (improvement > 1%) has been produces for 10 generations, then terminate.
Otherwise, continue with step 07.

07: **Ranking:** Individuals are ranked according to their fitness.

08: **Selection:** According to the size of the generational overlap select individuals for mating. High ranked individuals have a high probability to be selected.

09: **Mating:** Two-point crossover between a pair of selected individuals.

10: **Mutation:** Randomly change genes of individuals using a dynamic mutation probability scheme, with exponential decreasing probability during run-time. Mutation of communication mappings are randomly selected out of feasible assignments, depending on the task mapping.

11: **Offspring insertion:** Exchange low ranked individuals by newly produced individuals with respect to the size of the generational overlap. Continue with step 03.

Figure 4.14. Proposed **EE-GLSCMA** approach for combined optimisation of energy-efficient schedules and communication mappings

4.2.3 Experimental Results: Mapping Optimisation

The experimental results in this section analyse the effect of task and communication mapping on the achievable energy reductions through DVS. To evaluate this influence three basic concepts are compared:

CSE (Constructive Scheduling with fixed power DVS):
Within this approach, DVS is tackled under the fixed power model (i.e., no power variations are considered). Scheduling is carried out using a mobility-based constructive list scheduling. Communication mapping is based on a heuristic method that assigns communications to the first available CL. The task mapping is carried out using the genetic task mapping algorithm of Section 4.2.1.

DLSP (Dynamic Level Scheduling with power variation DVS):
DLSP is based on a constructive list scheduling which dynamically re-calculates task priorities during the schedule construction [138]. The priorities of tasks are given by the dynamic level, which depends on the longest path of the activity and the earliest start time. The priorities are used for a combined task mapping and scheduling. Thus, this approach is thoroughly of constructive nature. This scheduling and mapping approach corresponds to the one used in [20]. The so produced schedules are scaled using the PV-DVS algorithm introduced in Chapter 3.

ICMSP (Iterative Combined Mapping and Scheduling with PV-DVS):
This approach corresponds to the techniques and algorithms proposed in this book. That is, voltage scaling is performed by PV-DVS (Section 3.2.1), which considers the power variation model. Scheduling and communication mapping are based on the combined genetic algorithm EE-GLSCMA (Section 4.2.2). The task mapping is optimised using the genetic task mapping algorithm EE-GTMA (Section 4.2.1).

To ease the following discussion, Figure 4.15 provides a short overview of these optimisation concepts for reference purpose. All experimental results presented in this section are based on the same four benchmark sets (a–d) that have already been introduced in Sections 3.3 and 4.1. The presented results were obtained by running the optimisation process ten times and averaging the outcomes. The experiments are subdivided into two sections. Section 4.2.3.1 compares the CSE approach with ICMSP technique, while Section 4.2.3.2 assesses the optimisation potentials of DLSP with respect to ICMSP.

4.2.3.1 Comparison between CSE and ICMSP

The first experiments give a comparison between the CSE approach and the ICMSP approach. Table 4.5 shows this comparison for the benchmark sets

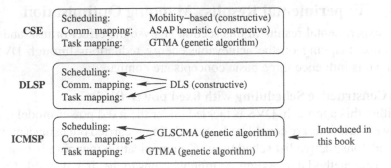

CSE	Scheduling:	Mobility–based (constructive)
	Comm. mapping:	ASAP heuristic (constructive)
	Task mapping:	GTMA (genetic algorithm)

DLSP	Scheduling:	
	Comm. mapping:	DLS (constructive)
	Task mapping:	

ICMSP	Scheduling:	GLSCMA (genetic algorithm) ← Introduced in this book
	Comm. mapping:	
	Task mapping:	GTMA (genetic algorithm)

Figure 4.15.   Three scheduling and mapping concepts

	NO-DVS		CSE			ICMSP (introduced)			
Example	*Energy* (μJ)	*CPU time (s)*	*Energy* (μJ)	*CPU time (s)*	*Reduc. (%)*	*Energy* (μJ)	*CPU time (s)*	*Reduc. (%)*	*Reduc. Factor*
tgff1*	333	3.11	116	1.91	65.23	92	12.14	72.41	1.11
tgff2	709747	24.10	625970	13.34	11.80	445532	47.86	37.23	3.15
tgff3	298991	69.46	225433	39.68	24.60	109351	2437.98	63.43	2.58
tgff4	63924	24.15	15743	12.15	75.37	10817	290.10	83.08	1.10
tgff4_t	49807	22.93	20275	11.83	59.29	18487	226.93	62.88	1.06
tgff4_f.	59294	20.32	18860	11.66	68.19	10621	299.45	82.09	1.20
tgff5	568210	64.42	426614	41.23	24.92	233063	904.99	58.98	2.37
tgff6	24685	19.97	7298	11.66	70.44	3799	221.99	84.61	1.2
tgff7	1491203	10.29	1169258	5.55	21.59	1058346	41.75	29.03	1.34
tgff8	525250	15.52	182894	8.49	65.18	136057	46.92	74.1	1.35
tgff9*	600428	9.01	358087	4.63	40.36	323158	45.28	46.18	1.14
tgff10	9417	7.45	8531	3.98	9.41	7193	17.58	23.62	2.51
tgff11	2858919	26.87	2400940	14.48	16.02	2229397	97.6	22.02	1.37
tgff12	174440	56.36	90087	37.61	48.36	58404	1328.52	66.52	1.38
tgff13	927704	60.97	511019	32.56	44.92	328377	853.42	64.6	1.44
tgff14	7723	23.29	7578	14.39	1.88	6693	69.01	13.34	7.11
tgff15	20017	86.85	17948	54.39	10.34	16938	916.98	15.38	1.49
tgff16	2984716	34.66	2177495	24.64	27.05	2141352	197.41	28.26	1.04
tgff17	16237	41.97	11417	26.80	29.69	8220	308.54	49.38	1.66
tgff18	1518517	4.17	1248236	2.80	17.80	1066350	9.71	29.78	1.67
tgff19	3431	5.91	2176	4.12	36.59	1907	18.31	44.41	1.21
tgff20*	18621	12.41	7286	7.18	60.87	4646	92.24	75.05	1.23
tgff21	2182722	121.95	1543090	59.71	29.30	1352422	1665.04	38.04	1.3
tgff22	894765	301.48	691269	172.38	22.74	456021	2240.57	49.03	2.16
tgff23*	5519226	147.33	3261600	87.85	40.90	2129198	14050.26	61.42	1.5
tgff24	720861	151.80	302288	98.07	58.07	200328	2199.39	72.21	1.24
tgff25	3232360	74.20	2555077	44.36	20.95	2328983	1664.63	27.95	1.33
Hou*	11816	10.57	10704	11.43	9.41	6708	163.78	43.23	4.59
Hou_c.*	12766	1.58	10145	1.97	20.53	7879	3.42	38.28	1.86

Table 4.5.   Mapping optimisation with and without DVS optimised scheduling using `tgff` and hou benchmarks

`tgff` and `Hou`. The first column gives the benchmark names. The second column shows the nominal energy dissipations. These values were produced using the genetic task mapping algorithm (**EE-GTMA**) in combination with the genetic scheduling and communication mapping algorithm (**EE-GLSCMA**), however, *without* using dynamic voltage scaling. Thus, all tasks are executed at nominal supply voltage. The third column gives the computational times for these optimisations. The nominal energy values are used as base for a comparison of the CSE and ICMSP approaches. The achieved results of the CSE approach are shown in Columns 4–6, while the outcomes of ICMSP are given in Columns 7–9. The last column provides the achieved reduction factors, comparing CSE and ICMSP. Consider, for instance, the benchmark example tgff4, which has a nominal energy dissipation of $63924\mu J$. This solution was found after optimising the application mapping and scheduling for $24.15s$. The energy dissipation of the same example is reduced to $15743\mu J$ using the CSE approach, a reduction by 75.37%. This has been achieved in smaller run-time, although the CSE additionally account for DVS. The lower optimisation time can be explained by the fact that a quick constructive scheduling technique is used. Furthermore, EVEN-DVS (fixed power model) does not significantly increase the optimisation time since it can be performed with linear computational complexity. Nevertheless, it is possible to further increase the energy savings using the ICMSP approach, which considers the power variation model as well as an iterative optimisation of the scheduling and mapping towards DVS.

Using ICMSP the energy consumption of example tgff4 is reduced to $10817\mu J$, a reduction by 83.08%. Due to the iterative scheduling and the quadratic complexity of PV-DVS, the optimisation time increases to $290.10s$. Comparing the achieved reductions of the constructive-based CSE approach (75.37%) with the introduced ICMSP approach (83.08%), a reduction factor of $83.08/75.37 = 1.10$ can be calculated. Observing the remaining reduction factors given in Table 4.5, it can be seen that ICMSP (based on the technique proposed in this book) always results in lower energy dissipations than the CSE approach (based on traditional scheduling techniques and the fixed power model). These reductions are mainly achieved due to two reasons:

(a) The schedule solution space can be more thoroughly search by an iterative scheduling technique.

(b) Considering the power variations allows a more accurate energy estimation to guide the optimisation process.

The run-times for the CSE technique varied between $1.91s$ (tgff1) and $172.38s$ (tgff22) for task graphs with up to 100 nodes. The ICMSP approach optimised all the examples in $3.42s$ (hou_clust) to $14050.26s$ (tgff22). Clearly, a trade-off between run-time and quality.

Timing Behaviour Improvements
In addition to the schedule discussions given in Section 4.1, the following observations about the scheduling optimisation can be made. The scheduling optimisation does not only significantly reduce the dissipated energy, but also

improves the timing behaviour, leading to feasible implementations where constructive techniques might fail. This is of great importance since high quality
solutions are likely to be found in design space regions where infeasible and
feasible solutions are spatially placed closely together. Making a wrong decision might involve a more costly implementation of the system. To clarify
this, consider the results obtained with the benchmark set TG1 from Gruian *et
al.* [68], as shown in Table 4.6. Scheduling the system tasks based on a con-

Example	NO-DVS		CSE			ICMSP (introduced)			
	Energy Dissip.	CPU time (s)	Energy Dissip.	CPU time (s)	Reduc. (%)	Energy Dissip.	CPU time (s)	Reduc. (%)	Reduc. Fac.
r000	798700	53.87	unsolved	18.60	n/a	586806	194.86	26.53	n/a
r001	759500	56.16	592674	13.87	21.97	399839	804.73	47.35	2.16
r002	744800	55.64	unsolved	16.51	n/a	551944	189.97	25.89	n/a
r003	994700	27.76	711887	15.98	28.43	554171	769.58	44.29	1.56
r004	886900	54.00	unsolved	19.97	n/a	566263	360.58	36.15	n/a
r005	744800	54.94	465853	16.75	37.45	373677	1596.67	49.83	1.33
r006	901600	36.88	unsolved	17.55	n/a	589469	827.22	34.62	n/a
r007	837900	55.20	unsolved	20.20	n/a	565731	269.07	32.48	n/a
r008	862400	30.63	unsolved	19.25	n/a	635426	207.46	26.32	n/a
r009	681100	53.24	424723	14.99	37.64	311751	1535.28	54.23	1.44
r010	705600	55.96	464257	17.88	34.20	325572	1155.62	53.86	1.57
r011	803600	32.11	558165	17.98	30.54	432977	421.37	46.12	1.51
r012	994700	49.54	unsolved	17.86	n/a	703960	310.65	29.23	n/a
r013	818300	45.58	unsolved	21.36	n/a	546724	315.27	33.19	n/a
r014	872200	56.16	618498	17.99	29.09	467035	2324.70	46.45	1.60
r015	813400	54.97	501944	17.12	38.29	427297	2891.08	47.47	1.24
r016	744800	26.52	unsolved	20.82	n/a	599931	141.81	19.45	n/a
r017	886900	56.54	unsolved	17.75	n/a	625210	121.10	29.51	n/a
r018	700700	54.81	unsolved	14.05	n/a	420155	581.39	40.04	n/a
r019	872200	28.29	unsolved	16.13	n/a	612864	172.78	29.73	n/a
r020	744800	27.92	unsolved	18.73	n/a	442314	357.80	40.61	n/a
r021	984900	53.81	726542	18.88	26.23	660399	2542.27	32.95	1.26
r022	818300	34.35	unsolved	20.66	n/a	467490	636.25	42.87	n/a
r023	798700	55.24	487590	17.64	38.95	353875	1455.71	55.69	1.43
r024	828100	54.25	unsolved	20.78	n/a	521388	455.70	37.04	n/a
r025	764400	55.28	378677	16.10	50.46	284444	2877.87	62.79	1.24
r026	877100	55.98	unsolved	16.02	n/a	624911	255.15	28.75	n/a
r027	715400	55.26	unsolved	14.18	n/a	483865	225.51	32.36	n/a
r028	813400	51.26	585257	18.38	28.05	443074	932.38	45.53	1.62
r029	656600	27.86	unsolved	19.08	n/a	439722	218.24	33.03	n/a

Table 4.6. Mapping optimisation of the benchmark set TG1 using NO-DVS (Nominal), EVEN-
DVS, and PV-DVS

structive list scheduling heuristic (mobility-driven) produces a single solution,
which might be feasible or infeasible. Consider, for example, benchmark r000.
In the case of this benchmark the constructive scheduling attempt fails and the
implementation is marked infeasible (Column 4, "unsolved"). Thus, making
it necessary to increase the performance of the allocated system for the given
mapping. On the other hand, the proposed iterative GA-based list scheduling
technique is able to improve infeasible schedules by providing feedback to the
optimisation process and therefore feasible schedules might be found, as in the

case of the task graph example r000 (Column 7). This effect is likely to appear in the presence of tight deadline specifications, such as the benchmark set TG1. It can be observed that for 18 out of 30 examples no feasible mapping could be found when using a mobility-driven scheduling algorithm. Nevertheless, using the genetic list scheduling approach it was possible to find feasible mappings for all task graphs of TG1, with energy reductions of up to 62.79% (benchmark r025). Compared to the feasible mappings generated using the CSE technique, energy reductions of up to 7.11 times could be achieved. This higher quality results require longer optimisation times.

4.2.3.2   Comparison between DLSP and ICMSP

To further confirm the ability of the proposed approach to optimise the mapping as well as the scheduling towards an effective utilisation of DVS, it is compared next with mappings and schedules produced by a dynamic level scheduling algorithm (DLS) [138]. This scheduling technique can be considered to be more sophisticated than "simple" mobility list scheduling approaches, since it dynamically re-calculates the task priorities after each schedule decision. Table 4.7 gives this comparison between DLSP and ICMSP. The first

Example	NO-DVS	DLSP		ICMSP (introduced)			
	Energy Dissip.	Energy Dissip.	Reduc. (%)	Energy Dissip.	CPU time (s)	Reduc. (%)	Reduc. Factor
fft1	29600	18154	38.67	14019	591.2	52.63	1.36
fft3	48000	36772	23.39	21452	1144.7	55.31	2.37
karp10	59400	47737	19.63	24055	755.1	59.50	3.07
meas	28300	25732	9.07	25732	8.5	9.07	1
qmf4	16000	12740	20.38	11097	202.3	30.64	1.50

Table 4.7. Comparison between DLS algorithm and the proposed scheduling and mapping approach using Bambha's benchmarks [20]

and the second column show the benchmark names and the nominal energy consumption, respectively. The third and fourth column represent the results obtained by the DLSP [3], while the fifth and the sixth column correspond to the proposed ICMSP approach. Accordingly, column seven gives the reduction factors when comparing DLSP with ICMSP. Examining the results, it can be seen that ICMSP was able to reduce the dissipated energy of 4 out of 5 examples by up to 39.87% ($59.50\% - 19.63\%$, karp10) equivalent to a reduction factor of 3.07. Only in the case of meas the dissipated energy remained the constant. This can be explained by the highly serialised graph structure of this task graph, which constrains the scheduling order and hence the potential to optimise the schedule. Furthermore, the serialised execution restricts the application parallelism and the used DVS-PEs are of identical types, thus, the task mapping does not influence the scheduling results. Nevertheless, for fft1, fft3, karp10, and qmf4, which show more parallelism, the introduced synthesis approach (ICMSP) outperforms the DLS-based scheduling in terms of energy savings by up to 3.07 times (karp10: 19.63% compared to 59.50%). It should be noted, both DLSP as well as ICMSP use dynamic voltage scaling under the power variation model. Hence, the reported saving are solely introduced by an improve application mapping and activity scheduling. Certainly, due the constructive nature of DLS, it surpasses the iterative improvement ICMSP approach in terms of optimisation time. For all examples the DLS run-time is below an optimisation time of $1s$, while the presented technique shows run-times between 8.5 and $1144.7s$. However, an advantage of the proposed approach is the possibility to initialise the mappings and schedules with a pre-passed mapping and scheduling based on DLS. This ensures that solutions of at least DLS quality are obtained in the first generation of the GA optimisation, that is, in an "identical" optimisation time. Such an initial solution pool could then be further optimised iteratively. In this way, the system designer can easily exploit the freedom to trade off synthesis time and solution quality.

Summary

This section has analysed the effect of application mapping under different constellations of scheduling and voltage scaling techniques. The conducted experiments have shown that substantial energy saving can be achieved by the iterative techniques introduced in this book, when compared to constructive scheduling and mapping approaches [138, 158], which have been used in previous work on energy minimisation through DVS [20, 99]. Furthermore, the experiments indicate an advantage of the proposed techniques over constructive approaches also in terms of schedulability in the presence of tasks with tight deadlines. The results reinforce the importance of a thorough exploration of the mapping and scheduling solution space. Clearly, the cost for these better results is higher computational time.

4.3   Optimisation of Allocation

Sections 4.1 and 4.2 have introduced techniques and algorithms for the optimisation of activity scheduling and application mapping, respectively. As outlined in Section 1.3, the overall goal of the co-synthesis process is to support the designer in finding the "most" suitable target architecture, i.e., the optimisation of the architecture allocation. In the proposed system-level design approach, this step is user-driven and thereby based on the knowledge and experience of the designer. It is assumed that the designer has predefined an architecture and the voltage scaling, scheduling, and mapping techniques help him to evaluate the quality of the allocation in terms of energy dissipation, cost, and feasibility. If an architecture does not prove to be satisfactory, the designer makes the necessary changes and evaluates again. In this way, it is also possible to trade off the different design goals and hence achieve multiple design alternatives. Similarly to the scheduling and mapping steps, the allocation of components has an influence on the usability of DVS. For example, it might be beneficial to reduce the workload on the system PEs by introducing a new PE or by re-allocating faster PEs. And thereby it could be possible to increase the deadline slacks in the system schedules and hence exploit them using DVS, resulting in higher dynamic energy reductions, while increasing the product cost and the static power consumption. Clearly, this optimisation is based on the astuteness of the designer. The following experiment demonstrates the importance of the architecture allocation using a real-world example.

4.3.1   Experimental Results: Component Allocation

In order to assess the energy reduction capability of the proposed synthesis approach in terms of real-world applicability, this set of experiments are concerned with a real-life optical flow detection (OFD) application. This application is a sub-system of an autonomous model helicopter for traffic monitoring purpose [12, 68], and it consists of 32 tasks. In its current implementation the OFD algorithm runs on two ADSP-21061L DSPs, with an average current of $760mA$ at $3.3V$, hence, an average power dissipation of approximately $2.52W$. Due to the stringent power budget on board of the helicopter, including application critical sub-systems, it is necessary to keep the overall power dissipation under a certain limit. To reduce the power consumption to a minimal amount, DVS seems predestined, since the OFD algorithm shows an unnecessary high performance (12.5 frames of 78x120 pixels per second). However, a repetition rate of 6.25 frames per second is sufficient (at certain operational heights) to ensure correct flow detection, allowing to relax the system constraints. For experimental purpose, a hypothetical extension of the DSPs towards DVS capability is considered. It is taken into account that such an extension has an influence on the static power dissipated by the digital circuits, and therefore the static power is increased by 10%.

In the first part of this experiment the application constraints are kept fixed, i.e., the OFD algorithm needs to perform with a repetition rate of $12.5Hz$ (equivalent to the current implementation). In order to increase the usage of the application parallelism, three different architectures are used, which are built out of three to five DVS-DSPs and connected via a shared bus. In this way the OFD algorithm can be performed faster, i.e., additional system slack is introduced, which is exploitable by DVS. Table 4.8 reports on the findings. The first row represents the current implementation of the OFD algorithm, i.e., running on an architecture consisting of two DSPs without DVS technology. This implementation shows a total average power dissipation of $2.52W$. Now, consider the architectures with three to five DVS-enabled DSPs. In accordance

Architecture	Static Power (W)	Dyn. Power (W)	Total Power (W)	CPU Time (s)	Reduc. (%)
2 DSPs (NO-DVS)	0.383	2.137	2.520	n/a	n/a
3 DVS-DSPs	0.574	1.371	1.945	148.3	22.8
4 DVS-DSPs	0.736	1.163	1.899	303.6	24.6
5 DVS-DSPs	0.898	1.132	2.030	381.9	19.4

Table 4.8. Increasing architectural parallelism to allow voltage scaling of the OFD algorithm

to the number of allocated PEs, the static power consumption increases. However, the increased number of PEs allows to exploit the application parallelism more effectively, which, in turn, allows a faster execution of the OFD algorithm. This results in slack time, usable by the DVS-PEs to lower the dynamic power dissipation. As it can be observed from Table 4.8, all implementations using DVS-DSPs show a reduction in total power consumption (sum of static and dynamic power consumption) of up to 24.6%. Note that this reduction does *not* necessitate any performance degradation, while the cost of the system increases.

As mentioned before, the current implementation of the OFD algorithm shows an unnecessary high performance and it is therefore possible to relax the system constraints. Hence, in the following experiment, the repetition rate is reduced from $12.5Hz$ to $6.25Hz$, i.e., an execution at half speed. This performance is still high enough to allow a correct flow detection. The results of this investigation are shown in Table 4.9. Observing the results shows that

Architecture	Static Power (W)	Dyn. Power (W)	Total Power (W)	CPU Time (s)	Reduc. (%)
2 DSPs (NO-DVS)	0.383	1.069	1.452	n/a	n/a
2 DVS-DSPs	0.413	0.394	0.807	783.5	44.4
3 DVS-DSPs	0.574	0.277	0.851	1107.2	41.4
4 DVS-DSPs	0.736	0.253	0.989	1393.4	31.9
5 DVS-DSPs	0.898	0.241	1.139	1634.7	21.6

Table 4.9. Relaxing the performance constraints of the OFD algorithm

for all given architectures the power dissipation could be reduced significantly, by up to 44.4% when compared to a non-DVS implementation (first row in Table 4.9). It is interesting to observe that the power consumption of the nominal task execution is reduced as well. This is due to the fact that the two DSPs are considered to consume no dynamic power when no computations are performed (through clock-gating). Therefore, for half of the operational time the DSPs dissipate no dynamic power. However, among all implementation alternatives

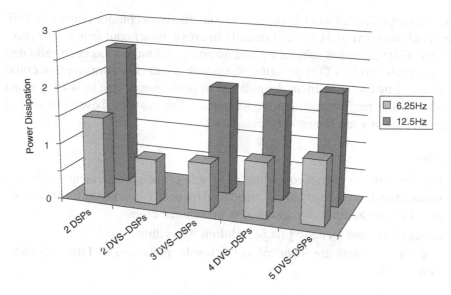

Figure 4.16. Nine different implementation possibilities of the OFD algorithm

the architecture composed out of two DVS-PEs (second row in Table 4.9) is the favourite, since it achieves the highest energy savings at a low cost. Its favourite position comes from the fact that with each additionally allocated PE the static power consumption increases, while the achievable dynamic energy reductions decrease (caused by limited parallelism within the application). Again, this shows how important an accurate design space exploration is when synthesising DVS-enabled embedded systems.

The power dissipations of all synthesised OFD systems are additionally shown in Figure 4.16. The depicted energy values correspond to the total energy dissipation, i.e., static as well as dynamic power dissipation are considered. Observing this chart clearly indicates the advantage of an OFD system using a two DVS-DSP implementation at a reduced performance rate of 6.25Hz.

4.4 Concluding Remarks

This chapter has introduced new methods for the activity scheduling and the application mapping of energy-efficient distributed heterogeneous embedded systems. A novel two-step iterative co-synthesis approaches that separates task mapping from a combined scheduling and communication mapping optimisation has been developed and applied to a set of benchmarks examples. By separating the two optimisation, the search space is pruned to structurally feasible solutions only, hence, potentially reducing the required synthesis times. It has been shown that the introduced methods not only achieve substantially higher energy savings when compared to constructive techniques (up to 39.87% for benchmark karp10), but additionally improve the schedulability. The practicality of the proposed scheduling and mapping techniques has been validated using a real-world OFD application, for which the energy consumption could be reduced by 24.6% without degradation in performance and by 44.4% when a reduction in performance can be tolerated. The cost for these solutions of improved quality are longer synthesis times.

Notes

1 For distributed systems consisting of several processing elements and communication links, a ready list is introduced for each component. Furthermore, priorities are additionally assigned to communications.
2 Energy-Efficient Genetic List Scheduling Algorithm
3 The DLSP results are identical to the results presented in Table 3.5 (Section 3.3.1).

Chapter 5

ENERGY-EFFICIENT MULTI-MODE EMBEDDED SYSTEMS

Chapter 3 has shown how the power variation model helps to reduce the energy consumption of DVS-enabled embedded systems during voltage scaling. Furthermore, it was shown in Chapter 4 how an appropriate application mapping and activity scheduling enables an effective utilisation of DVS, in order to save energy. These techniques and algorithms have addressed the optimisation of embedded systems that perform one single application, such as a standalone MP3 decoder or an optical flow detection (OFD). Nevertheless, one key characteristic of many current and emerging embedded systems is their need to work across a set of different interacting applications and operational modes. For instance, modern mobile phones often integrate not solely the functionality required for communication purpose, but additionally integrate applications like digital cameras, games, and complex multimedia functions (MP3 players and video decoders) into the same single device. In this book, such embedded systems are referred to as *multi-mode embedded systems*. This chapter introduces a novel co-synthesis methodology for the design of energy-efficient multi-mode embedded systems. Starting from a specification model that captures both mode interaction and functionality, the developed co-synthesis technique maps the application under consideration of *mode execution probabilities* to a heterogeneous architecture with the aim to reduce the energy consumption through an appropriate resource sharing between tasks. The main principle by which the co-synthesis methodology achieves energy-efficiency is an implementation trade-off between the different operational modes. In general, modes with high execution probability should be implemented more energy efficient (e.g., by moving more tasks to hardware) than modes with a low execution probability. Nonetheless, the implementation of modes is heavily interrelated, due to the fact that different modes share the same resources (architecture). For example, mapping an energy-critical task of a highly active mode into energy-efficient hardware might prohibit to implement a timing-critical task into hardware due to the restricted hardware area (see motivational example in Section 5.2). Clearly, a well balanced implementation of the operational modes is vital for a good system design. In addition, the co-synthesis approach further reduces the energy dissipation by adapting the system performance to the particular needs of the active mode, using dynamic voltage scaling as well as component shutdown. Furthermore, the voltage scaling method of Chapter 3 is extended to account for hardware PEs that are capable of executing tasks in parallel, however, rely on a single scalable supply voltage source.

The rest of this chapter is organised as follows. Preliminaries, regarding the specification and the architectural models are given in Section 5.1. In Section 5.2, the problem addressed in this chapter is motivated through illustrative examples. Section 5.3 surveys relevant previous work regarding multi-mode embedded system. A formulation of the problems at hand is provided in Section 5.4. Section 5.5 proposes a novel synthesis approach, in order to tackle the identified problems. Extensive experimental results, including a smart phone example, are given in Section 5.6. Finally, concluding remarks are expressed in Section 5.7.

5.1   Preliminaries

This section introduces the functional specification model (Section 5.1.1) and the architectural model (Section 5.1.2), which are fundamental to the co-synthesis framework outlined in this chapter.

5.1.1   Functional Specification of Multi-Mode Systems

The abstract model used for the specification of multi-mode embedded systems consists of two parts. In précis, it is based on a combination of finite state machines and task graphs, capturing both the interaction between different operational modes as well as the functionality of each individual mode. Structurally, each node in the finite state machine represents an operational mode and further contains the task graphs which are active during this mode. The following two sections introduce this model, which is henceforth referred to as operational mode state machine (OMSM).

Top-level Finite State Machine

In this chapter, it is considered that an application is given as a directed cyclic graph $\Upsilon(\Omega, \Theta)$, which represents a finite state machine. Within this top-level model, each node $O \in \Omega$ refers to an operational mode and each edge $T \in \Theta$ specifies a possible transition between two different modes. If the system undergoes a change from mode O_x to mode O_y, where $x \neq y$, the transition time t_T^{max} associated with the transition edge $T = (O_x, O_y)$ has to be met. At

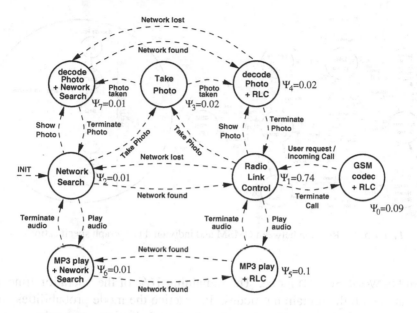

Figure 5.1. Example operational mode state machine of a smart phone

any given time there is only one active mode, i.e., the modes execute mutually exclusive. To exemplify the proposed model consider Figure 5.1. This figure shows the operational mode state machine for a smart phone example with eight different modes. A possible activation scenario could look like this: When switched on, the phone initialises into **Network Search** mode. The system stays in this mode until a suitable network has been found. Upon finding a network the phone undergoes a mode change to **Radio Link Control** **(RLC)**. In this mode it maintains the connection to the network by handling cell hand-overs, radio link failure responses, and adaptive RF power control. An incoming phone call necessitates to switch the system into **GSM codec +** **RLC** mode. This mode is responsible for speech encoding and decoding, while simultaneously maintaining network connectivity. Similarly, the remaining modes have different functionalities and are activated upon mode change events. Such events originate upon user requests (e.g. MP3-player activation) or are initiated by the system itself (e.g. loss of network connection necessitates to switch the system into **network search** mode). Furthermore, based on the key observation that many multi-mode systems spend their operational time *unevenly* in each of the modes, an execution probability Ψ_O is associated with each operational mode O, i.e., it is known what percentage of the operational time the device spends in each mode. For instance, in accordance to the typical values given in Figure 5.1, the smart-phone stays 74% of this operational time in **Radio Link Control (RLC)** mode, 9% in **GSM codec + RLC** mode, and

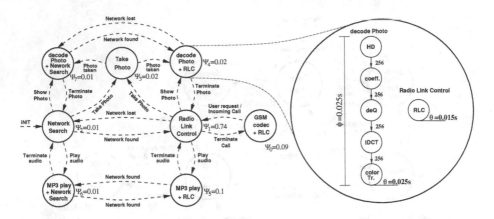

Figure 5.2.   Relation between OMSM and individual task graph specifications

1% in **Network Search** mode. The remaining 16% of the operation time are
associated with the remaining modes. In practice the mode probabilities vary
from user to user, depending on the personal usage behaviour. Nevertheless, it is
possible to derive an average activation profile based on statistical information
collected from several different users. Taking this information into account
will prove to be important when designing systems with a prolonged battery
lifetime.

It is interesting to note that different operational modes do not necessarily
correspond to different functionalities of the system. For example, alternative
modes can be used to model the same functionality under different working
conditions (such as different workloads). For instance, in order to account
for variations in the wireless channel quality, we could exchange the GSM
voice transcoder mode O_0 in Figure 5.1(a) with three transcoder schemes, each
responsible for the coding at a different signal-to-interference ratio (SIR) on
the channel. During run-time the appropriate transcoder scheme would be
selectively activated, depending on the actual channel quality.

Functional Specification of Individual Modes

The functional specification of each operational mode $O \in \Omega$ in the top-level finite state machine is expressed by a task graph $G_S^O(\mathcal{T}, \mathcal{C})$. This relation is shown
in Figure 5.2. The task graph model was introduced in Chapter 1, Section 1.2.1.
However, due to some particularities concerning multi-mode systems, the following outlines the exact model: Each node $\tau \in \mathcal{T}_O$ in a task graph represents a
task, i.e., an atomic unit of functionality that needs to be executed without pre-
emption. The level of granularity is coarse, i.e., tasks refer to functions such as
Huffman decoder, de-quantizer, FFT, IDCT, etc. Therefore, each task is further
associated with a task type $\eta \in \Gamma = \{HD, deQ, FFT, IDCT, ...\}$. A distinc-

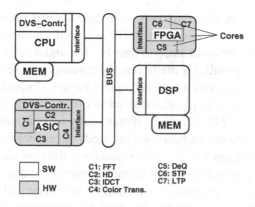

Figure 5.3. Distributed Architectural Model

tive feature of multi-mode systems is that task type sets $\Gamma^O \subseteq \Gamma$ of different modes $O \in \Omega$ can intersect, i.e., tasks of identical type can share the same hardware resource (inter-mode sharing). Resource sharing is also possible for multiple tasks of identical type that are found in a single mode (intra-mode sharing), however, due to task communalities among different modes, the chances to share resources are increased. Edges $\gamma \in C$ in the task graph refer to precedence constraints and data dependencies between the computational tasks, i.e., if two tasks, τ_i and τ_j, are connected by an edge, then task τ_i must be finished and transfer data to task τ_j, before τ_j can be executed. A feasible implementation of a single mode O needs to respect all task deadlines θ, task graph period ϕ, and precedence relations.

5.1.2 Architectural Model and System Implementation

Similar to the techniques introduced in the Chapters 3 and 4, the proposed system-level synthesis approach targets distributed architectures that possibly consist of several heterogeneous processing elements (PEs), such as general-purpose processors (GPPs), ASIPs, ASICs, and FPGAs. These components are connected through an infrastructure of communication links (CLs). A directed graph $G_A(\mathcal{P}, \mathcal{L})$ captures such an architecture, where nodes $\pi \in \mathcal{P}$ and edges $\lambda \in \mathcal{L}$ denote PEs and CLs, respectively. Figure 5.3 shows an architecture example. Since each task might have multiple implementation alternatives, it can be potentially mapped onto several different PEs that are capable of performing this type of task. Tasks mapped to software-programmable components (i.e., GPP or ASIP) are placed into local memory. However, if a task is mapped to a hardware component (i.e., ASIC or FPGA), a core for this task type needs to be allocated. A feasible solution needs to obey the imposed area constraints, i.e., only a restricted number of cores can be implemented on hardware components. The subdivision of hardware components (ASICs and FPGAs) into hardware cores is shown in Figure 5.3. Each core is capable of performing a single task of type $\eta \in \Gamma$ at a time. Tasks assigned to GPPs or ASIPs (software tasks) need to be sequenced, whilst the tasks mapped onto FPGAs and ASICs (hardware tasks) can be performed in parallel if the necessary resources (cores) are not already engaged. However, contention between two or more tasks assigned to the same hardware core requires a sequential execution order, similar to software tasks. Cores implemented on FPGAs can be dynamically reconfigured during a mode change, involving a time overhead, which needs to respect the imposed maximal mode transition times t_T^{max}. Further, similar to the problems investigated in Chapter 3 and Chapter 4, PEs might feature dynamic voltage scaling to enable a trade-off between power consumption and performance that can be exploited during run-time. A set $\mathcal{V}_\pi$ specifies the available discrete voltages of DVS-PE π. For such PEs a voltage schedule needs to be derived, in addition to a timing schedule. To implement a multi-mode application captured as OMSM, the tasks and communications of all operational modes need to be mapped onto the architecture, and a valid schedule for these activities $\epsilon \in \mathcal{A}$, where $\mathcal{A} = \mathcal{T} \cup \mathcal{C}$), needs to be constructed. Further, for tasks mapped to DVS-enabled components an energy reducing voltage schedule has to be determined. According to these aspects, an implementation candidate can be expressed through four functions, which need to be derived for each operational mode $O \in \Omega$:

Task mapping: $M_\tau^O : \mathcal{T} \to \mathcal{P}$

Communication mapping: $M_\gamma^O : \mathcal{C} \to \mathcal{L}$

Timing schedule: $S_\epsilon^O : \mathcal{A} \to \mathbb{R}_0^+$

Voltage schedule: $V_\tau^O : \mathcal{T}_{DVS} \to \mathcal{V}_\pi$

where M_τ^O and M_γ^O denote task and communication mapping, respectively, assigning tasks to PEs and communications to CLs. Activity start times are specified by the scheduling function S_ϵ^O, while V_τ^O defines the voltage schedule for all tasks $\tau \in \mathcal{T}_{DVS}$ mapped to DVS-PEs, where $\mathcal{V}_\pi$ is the set of the possible discrete supply voltages of PE π. Clearly, the mappings as well as the corresponding schedules are defined for every mode separately, i.e., during the change from mode O_x to mode O_y, the execution of activities found in mode O_x are finished, and the activities of mode O_y are activated.

5.2   Motivational Examples

The aim of this section is to motivate the key ideas behind the new multi-mode co-synthesis, that is, the consideration of mode execution probabilities and multiple task type implementations. First, the influence of mapping in the context of multi-mode embedded systems with different mode execution

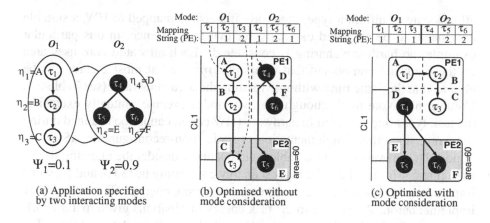

Figure 5.4. Mode execution probabilities

probabilities is demonstrated. Second, it is illustrated that multiple task implementations can help to reduce the energy dissipation of multi-mode embedded systems.

Example: Mode Execution Probabilities

For simplicity, timing and communication issues are neglected in the following example. Consider the application shown in Figure 5.4(a), which consists of two operational modes, O_1 and O_2, each specified by a task graph with three tasks. The system spends 10% of its operational time in mode O_1 and the remaining 90% in mode O_2, i.e., the execution probabilities are given by $\Psi_1 = 0.1$ and $\Psi_2 = 0.9$. The specification needs to be mapped onto a target architecture built of one general-purpose processor (PE1) and one ASIC (PE2), linked by a bus (CL1). Depending on the task mapping to either of the components, the execution properties of each task are shown in Table 5.1. It can be observed that

task type	PE1 (SW)		PE2 (HW)		
	exec. time (ms)	dyn. energy (mJ)	exec. time (ms)	dyn. energy (mJ)	area (mm^2)
A	20	10	2	0.010	24.0
B	28	14	2.2	0.012	30.0
C	32	16	1.6	0.023	27.5
D	26	13	3.1	0.047	24.5
E	30	15	1.8	0.015	21.0
F	24	14	2.2	0.032	28.0

Table 5.1. Task execution and implementation properties

all tasks are of different type, therefore, if a task is mapped to HW, a suitable core needs to be allocated explicitly for that task. Hence, in this particular example, no hardware sharing is considered. Each allocated core uses area on the hardware component that offers $60mm^2$, i.e., at most 2 cores can be allocated at the same time without violating the area constraint (see Table 5.1, Column 6). Note that although the two modes execute mutually exclusive, the task types implemented in hardware (HW cores) cannot be changed during run-time, since their implementation is static (non-reconfigurable ASIC); as opposed to software-programmable components. Consider the mapping shown in Figure 5.4(b), in which the highest energy consuming tasks (τ_3 and τ_5, when implemented in software) are executed using a more energy-efficient hardware implementation. According to the task energy dissipations given in Table 5.1, the energy dissipation during modes O_1 and O_2 are $E_1 = 10mJ + 14mJ + 0.023mJ = 24.023mJ$ and $E_2 = 13mJ + 0.015mJ + 14mJ = 27.015mJ$. Neglecting the mode execution probabilities by assuming that both modes are active for even amounts of time (50% mode 0_1 and 50% mode O_2) energy consumption can be calculated as $E_e = 0.5 \cdot 24.023mJ + 0.5 \cdot 27.015mJ = 25.519mJ$. Nevertheless, taking the real behaviour into account, mode O_1 is active for 10% of the operational time, i.e., its energy dissipation can then be calculated as $E_{r1} = 0.1 \cdot 24.023mJ = 2.4023mJ$. Similarly, mode O_2 is active 90% of the operational time, hence, its energy is given by $E_{r2} = 0.9 \cdot 27.015mJ = 24.3135mJ$. Based on both modes, the real energy dissipation results in $E_r = E_{r1} + E_{r2} = 26.7158mJ$. Now consider an alternative mapping for the same task graph, shown Figure 5.4(c). In this configuration tasks τ_5 and τ_6, i.e., the most energy dissipating tasks of the highly active mode O_2, use energy-efficient hardware implementations on PE2, while task τ_3 of the less active model O_1 is shifted into the software-programmable processor (PE1). According to this solution, the energy consumptions of modes O_1 and O_2 are given by $E_1 = 10mJ + 14mJ + 16mJ = 40mJ$ and $E_2 = 13mJ + 0.015mJ + 0.032mJ = 13.047mJ$. Considering the even execution of each mode (neglecting the execution probabilities), the even energy consumption can be calculated as $0.5 \cdot 40mJ + 0.5 \cdot 13.047mJ = 26.524mJ$. Note that this value is higher than the even energy of the first mapping ($E_e = 25.519mJ$). Thus, a co-synthesis approach that neglects the mode execution probabilities would optimise the system towards the first mapping. However, in real-life the modes are active for different amount of time and hence the real energy dissipation is given by $E_r = 0.1 \cdot 40mJ + 0.9 \cdot 13.047mJ = 15.7423mJ$. This is 41% lower compared to the first mapping ($E_r = 26.7158mJ$) shown in Figure 5.4(b), which is not optimised for an uneven task execution probability. Furthermore, the second task mapping allows to switch off PE2 and CL1 during mode O_1, since all tasks of this mode are assigned to PE1. This results in a

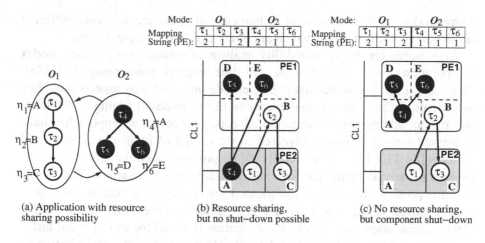

(a) Application with resource sharing possibility

(b) Resource sharing, but no shut-down possible

(c) No resource sharing, but component shut-down

Figure 5.5. Multiple task type implementations

significant reduction of the static power, additionally increasing the energy savings.

Example: Multiple Task Type Implementations

An important characteristic of multi-mode systems is that tasks of the same type might be found in different modes, i.e., resources can be shared among the different modes in a time-multiplexed fashion. To increase the possibility of component shutdown, it might be necessary to implement the same task type multiple times, however, on different components. The following example, shown in Figure 5.5, clarifies this aspect. Here tasks τ_1 and τ_4 are of type A (see Figure 5.5(a)), allowing resource sharing between these tasks. The sharing is possible without contention due to the mutual exclusive execution of these tasks, since only one mode is active at a given time. In the first mapping, given in Figure 5.5(b), both tasks utilise the same HW core. However, implementing task τ_4 in software (additional task type A on PE1), as shown in Figure 5.5(c), allows to shut down PE2 and CL1 during the execution of mode O_2. Hence, multiple implementations of task types can help to reduce power dissipation.

These two examples have demonstrated that it is essential to guide the synthesis process by: (a) an energy model that takes into account the mode execution probability as well as (b) allowing multiple task implementations.

5.3   Previous Work

Over the last few year, numerous methodologies for the design of low power consuming embedded systems have been proposed, including approaches that leverage power management techniques, such as dynamic power management (DPM) and dynamic voltage scaling (DVS). Nevertheless, a crucial feature of many modern embedded systems is their capability to execute several different applications (multi-modes), which are integrated into a single device.

Approaches for the schedulability analysis of systems with several modes of operations can be found in the real-time research community [117, 135]. However, these approaches solely concentrate on scheduling aspects (i.e., they investigate if the mode change events fulfil the imposed timing constraints) and do not address implementation aspects. Three recent approaches have addressed various problems involved in the design of multi-mode embedded systems [84, 113, 136]. Shin *et al.* [136] proposed a schedulability-driven performance analysis technique for real-time multi-mode systems. They show that it is possible, through a sophisticated performance estimation, to identify timing critical tasks, which are active in different operational modes. This identification allows to improve the execution times of the most crucial tasks, in order to achieve system schedulabilitly. In their work, the optimisation of the identified tasks is up to the designer. For example, reductions in the execution times can be made by handcrafted code tuning and outsourcing of core routines into hardware. Kalavade and Subrahmanyam [84] have introduced a hardware/software partitioning approach for systems that perform multiple functions. Their technique classifies tasks, found within similar applications, into groups of task types. The implementation of frequently appearing task types is biased towards hardware. This can be intuitively justified by the fact that costly hardware implementations are shared across a set of applications, hence, exploiting the allocated hardware more effective. Oh and Ha [113] address the problem in a slightly different way. Their co-synthesis framework for multi-mode systems is based on a combined scheduling and mapping technique for heterogeneous multiprocessor systems (HMP [111]). Taking a processor utilisation criterion into account, an allocation-controller selects the required processing elements such that the schedulability constraint is satisfied and the system cost is minimised. The main principle behind all three approaches is to consider the possibility of resource sharing, i.e., computational tasks of the same type, which can be found in different modes, utilise the same implementations. Thereby, multiple hardware implementations of the same task type are avoided, which, in turn, reduces the hardware cost. As opposed to these approaches, the work presented in this chapter addresses the design of low energy consuming multi-mode systems; hence, it differs in several aspects from the previous works. This chapter investigates the following subjects:

(a) The consideration of *mode execution probabilities* and their effect on the energy-efficiency of multi-mode embedded systems is analysed and demonstrated.

(b) A co-design methodology for the design of energy-efficient multi-mode systems is presented. The proposed co-synthesis maps and schedules a system specification that captures both mode interaction and mode functionality onto a distributed heterogeneous architecture. Four mutation strategies are introduced that aid the GA based optimisation process in finding solutions of high quality by pushing the search into promising design space regions.

(c) Dynamic voltage scaling is investigated in the context of multi-mode embedded systems. A transformation-based approach is used to tackle the problem of DVS on processing elements that execute different tasks in parallel, but that offer only a single scalable supply voltage source.

5.4   Problem Formulation

The goal of the introduced co-synthesis is an energy-efficient and feasible implementation of application Υ, modelled as OMSM. This involves the derivation of the mapping and schedule functions, M_{τ}^{O}, M_{γ}^{O}, S_{ϵ}^{O}, and V_{τ}^{O} (outlined Section 5.1.2), under the consideration of static and dynamic power as well as mode execution probabilities. Although the technique and algorithms introduced in the Chapters 3 and 4 concentrate on the minimisation of the dynamic power, in multi-mode systems static power consumption can have a significant impact on the overall energy efficiency. The reasons for this are the different performance requirements of the various operational modes. For instance, the minimal performance requirements of the hardware architecture are imposed by the most computational intensive mode, i.e., the minimal allocated architecture has to provide enough computation power to execute this performance critical mode. However, the allocated architecture might be far too powerful for the execution of modes with low performance needs. Furthermore, low performance modes, such as the standby-mode of mobile phones (i.e., Radio Link Control), often account for the greatest portion of the system time. During such circumstances, the static energy dissipation of unnecessarily switched-on PEs and CLs can outweigh the dynamic energy consumption caused by tasks of a "lightweight" mode. Thus, switching-off the unneeded components becomes an important aspect particularly in multi-mode embedded systems. In accordance, an accurate estimation of the average power consumption of an implementation alternative should consider both static and dynamic power, and furthermore the mode execution probabilities. The average power consumption $\bar{p}$ can be expressed using the following equation:

$$\bar{p} = \sum_{O \in \Omega} (P_O^{stat} + P_O^{dyn}) \cdot \Psi_O \qquad (5.1)$$

where P_O^{stat}, P_O^{dyn}, and Ψ_O refer to the static power dissipation, the dynamic power dissipation, and the execution probability of mode O, respectively. The static and dynamic power consumptions are given as:

$$P_O^{stat} = \sum_{\xi \in \mathcal{K}_O} P^{stat}(\xi) \tag{5.2}$$

and

$$P_O^{dyn} = \left(\sum_{\epsilon \in \mathcal{A}_O} E^{dyn}(\epsilon) \right) \cdot \frac{1}{hp_O} \tag{5.3}$$

where $P^{stat}(\xi)$ refers to the static power consumption of a component ξ, which is found in the set of all active components $\mathcal{K}_O \subseteq (\mathcal{P} \cup \mathcal{L})$ of mode O. Please note that this static power consumption also includes the additional power required for the DC/DC converter of voltage-scalable processors. Further, $\mathcal{A}_O$ and hp_O denote all activities and the hyper-period of mode O, respectively. With respect to the type of activities, the dynamic energy consumption $E^{dyn}(\epsilon)$ can be calculated in the same way as introduced in Equation (4.2) of Section 4:

$$E^{dyn}(\epsilon) = \begin{cases} P_{max}(\epsilon) \cdot t_{min}(\epsilon) \cdot \frac{V_{dd}^2(\epsilon)}{V_{max}^2(\epsilon)} & \text{if } \epsilon \in \mathcal{T}_{\text{DVS}} \\ P_{max}(\epsilon) \cdot t_{min}(\epsilon) & \text{if } \epsilon \in \mathcal{T} \setminus \mathcal{T}_{\text{DVS}} \\ P_C(\epsilon) \cdot t_C(\epsilon) & \text{if } \epsilon \in \mathcal{C} \end{cases} \tag{5.4}$$

where P_{max} is the dynamic power consumption and t_{min} the execution time of tasks when executed at nominal supply voltage V_{max}. Tasks $\tau \in \mathcal{T}_{\text{DVS}}$ mapped to DVS-PEs can execute at a scaled supply voltage V_{dd}, resulting in a reduced energy consumption. Further, communications consume power P_C over a time t_C. If the DVS-enabled processors are restricted to a limited set of discrete voltages, the continuous selected supply voltage V_{dd} is split into its two neighboring discrete voltages V_{low} and V_{high}. The corresponding execution times in each voltage are calculated as given in Section 3.2.2. The mode execution probabilities used in Equation (5.1) are either based on approximations or statistical information collected from several real users. In the case that statistical information is available from a set of different users U, the average execution probabilities $\overline{\Psi}_O$ of a single operational mode $O \in \Omega$ can be calculated.

The co-synthesis goal is to find a task mapping M_τ^O, a communication mapping M_γ^O, a starting time schedule S_ϵ^O, as well as a voltage schedule V_τ^O for each operational mode O, such that the total average power $\bar{p}$, given in Equation (5.1), is minimised. Furthermore, a feasible implementation candidate needs to fulfil the following requirements:

(a) The mapping of tasks M_τ^O does not violate area constraints in terms of memory and hardware area, i.e., $(\sum_{\eta \in \Gamma_\pi} a_\eta) \leq a_\pi^{max}$, $\forall \pi \in \mathcal{P}$; where Γ_π is the set of all task types implemented on PE π, and a_η and a_π^{max} refer to the area used by task type η and the available area on PE π, respectively.

Please note that for DVS-enabled HW, a_π^{max} represents the available area including the area overhead required for the DC/DC converter.

(b) The timing schedule S_ϵ^O and the voltage schedule V_τ^O, based on task and communication mapping, do not exceed any task deadlines θ_τ or task graph repetition periods ϕ, therefore, $t_S(\tau) + t_{exe}(\tau) \leq \min(\theta_\tau, \phi)$, $\forall \tau \in \mathcal{T}$; where $t_S(\tau)$ and t_{exe} refer to task start time and task execution time (potentially based on voltage scaling).

(c) The system reconfiguration time t_T between mode changes does not exceed the imposed maximal mode transition times t_T^{max}. Hence, $t_T \leq t_T^{max}$, $\forall T \in \Theta$ needs to be respected for all mode transitions.

5.5 Co-Synthesis of Energy-Efficient Multi-Mode Systems

Energy minimisation techniques for mapping and activities scheduling of single mode embedded systems have already been introduced in Chapter 4. This section proposes new techniques for the co-synthesis of energy-efficient multi-mode embedded systems. Similar to the algorithms given in Chapter 4, the co-synthesis for multi-mode systems is based on two nested optimisation loops. The outer loop optimises task mapping and core allocation, while the inner loop is responsible for the combined optimisation of communication mapping and scheduling. While the communication mapping and scheduling in the multi-mode co-synthesis approach are based on the algorithms given in Section 4.2.2, this section reconsiders task mapping, hardware core allocation, and dynamic voltage scaling to suit the particular problems of multi-mode embedded systems. Hence, the remainder of this section is organised as follows. We first discuss in Section 5.5.1 how mode execution probabilities be obtained in practice. Section 5.5.2 outlines a new co-synthesis algorithm for multi-mode systems, concentrating on task mapping and four improvement strategies that aid to tackle the problem of multi-mode task mapping. Section 5.5.3 outlines a heuristic technique for hardware core allocation. Finally, Section 5.5.4 describes a transformation-based approach for dynamic voltage scaling of parallel execution task on single hardware components.

5.5.1 Estimation of Mode Execution Probabilities

As we have demonstrated in the motivational example of Section 5.2, the consideration of mode execution probabilities during design time can help to significantly reduce the energy consumption of the embedded system. Certainly, to achieve a good design it is necessary that the execution probabilities (estimations) used during design time reflect the real usage probabilities (in-field) accurately. In the following, we outline how to obtain adequate execution probabilities using two different design scenarios:

(a) *The new design is an upgrade of an existing product which is connected to a service provider* (e.g. a new version of a mobile phone). For such product types it is possible to use information regarding the activation profile that has been collected on the provider side during the operation of the previous product generation. For instance, the cellular network base stations can record the activation profile of the mobile terminals (e.g. phones) regarding radio link control and calling mode, directly in-field, such as the one shown in Figure 5.6. The Figure gives the activation profile for a single phone during an operation period of 24 hours. According to this profile, the phone stays most of the time in a **Radio Link Control** mode, in order to maintain network connectivity. While the **Network Search** mode and the **Calling** mode are only active for small periods of the overall time. Using this information for a large number of phones could then be evaluated and used during the design of the new product.

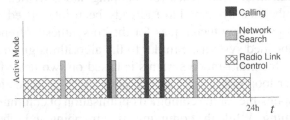

Figure 5.6.   Typical Activation Profile of a Mobile Phone

(b) *The product is a completely new design.* In this situation, it is common practice to evaluate the market acceptance before the final product is introduced using a limited number of prototypes that are distributed among a set of evaluation users. During this evaluation phase, the prototypes can gather information regarding the activation profile. This information could then be used during the final design of the product to optimize the energy consumption. Of course, it is also possible to use application-specific insight of the designer to estimate the execution probabilities. As we will show in the experiments given in Section 5.6.1, even if the estimated execution probabilities do not reflect the user activation with absolute accuracy, but are sufficiently close to the real values, energy savings can be still achieved.

5.5.2   Multi-Mode Co-Synthesis Algorithm

The task mapping approach, which determines M_T for all modes of application Υ is an enhancement of the genetic task mapping algorithm (**EE-GTMA**) introduced in Section 4.2.1. These enhancements include the consideration of resource sharing, component shutdown, and mode transition issues. As outlined in Section 4.1.2, GAs optimise a population of individuals over several

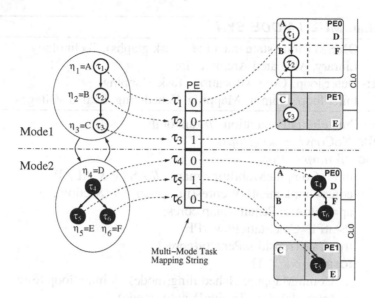

Figure 5.7. Task mapping string for multi-mode systems

generations by imitating and applying the principles of natural selection. That is, the GA iteratively evolves new populations by mating (crossover) the fittest individuals (highest quality) of the current population pool until a certain convergence criterion is met. In addition to mating, mutation, i.e., the random change of genes in the genome (string), provides the opportunity to push the optimisation into unexplored search space regions. As opposed to the single mode task mapping strings of Section 4.2, task mapping strings for multi-mode specifications combine the mapping strings of each operational mode into one large task mapping string as shown in Figure 5.7. Within this string each number represents the PE to which the corresponding task is assigned. This encoding enables the usage of a genetic algorithm to optimize the placement of tasks across the processing elements that form the distributed architecture. Please note that this representation supports the implementation of multiple task types. For instance, if two tasks of the same type are mapped onto different PEs, this tasks are implemented on both PEs. Thereby, the possibility of multiple task implementations is mainly inherited into the genetic mapping algorithm which is guided by a cost function that accounts for the multiple task implementations, i.e., the GA trades off between the savings in static power consumption against the increase dynamic power.

The goal of the co-synthesis is to find a mapping of tasks that minimises the total power consumption and obeys the performance constraints. Figure 5.8 outlines the pseudo-code of the co-synthesis algorithm. Starting from an initial random population of multi-mode task mapping strings (line 1), the optimi-

---

Algorithm: *MULTI-MODE-SYN*

Input: - OMSM (finite state machine + task graphs), Technology
               Library, Allocated Architecture
Output:- Outer loop: Core Allocation, Task Mapping
           - Inner loop: Comm. Mapping, Scheduling, Scaled Voltages

---

```
01: Pop=CreateInitialPopulation   // randomly
02: while(NoConvergence(Pop))
03:    forall map ∈ Pop
04:        mob=ComputeMobilities(map)   // ASAP & ALAP
05:        cores=ImplementHWcores(map,mob)   // (Section 5.5.3)
06:        ap=CalcAreaPenalty(map,cores)
07:        PstatPE=CalcStaticPowerPE
08:        trp=CalcTransitionPenalty(cores)
09:        forall mode ∈ Ω
10:            CommMapping_Scheduling(mode)   // inner loop (Sec. 4.2.2)
11:            tp(mode)=CalcTimingPenalty(mode)
12:            Pdyn(mode)=CalcDynPower(mode)   // incl. DVS
13:            PstatCL=CalcStaticPowerCL
14:        FM=MappingFitness(Pdyn,tp,PstatCL,PstatPE,ap,trp)
15:    ran=RankingIndividuals(FM)
16:    mat=SelectedMatingIndividuals(ran)
17:    TwoPointCrossover(mat)
18:    OffspringInseration(Pop)
19:    ShutdownImprovementMutation(Pop)
20:    AreaImprovementMutation(Pop)
21:    TimingImprovementMutation(Pop)
22:    TransistionImprovementMutation(Pop)
```

Figure 5.8.   Pseudo Code: Multi-Mode Co-Synthesis

sation runs until the convergency criterion is met (line 2). The used criterion is based on the diversity in the current population and the number of elapsed iterations without producing any improved individual. To judge the quality of mapping candidates, i.e., the fitness which guides the genetic algorithm, it is necessary to estimate important design objectives, including static and dynamic power dissipation, area usage, and timing behaviour (lines 03–13). The following explains each of the required estimations. The hardware area depends on the allocated cores on each hardware component (ASIC or FPGA). Of course, for each task type mapped to hardware at least one core of this type needs to be allocated. However, if too many cores are placed onto a single

ASIC or FPGA, the available area is exceeded and an area penalty is introduced (line 6). On the other hand, if multiple tasks of the same type are mapped to the same hardware component and the hardware area is not violated, it is possible to implement cores multiple times (if helpful for the energy reduction). In the proposed approach, additional cores (line 5) are allocated for parallel tasks with low mobility (line 4), therefore, the chance to exploit application parallelism is increased. Clearly, from an energy point of view, this is also preferable, especially in the presence of DVS, where a decreased execution time can be exploited. Section 5.5.3 describes the core allocation in more detail. At this point it is possible to compute the static power consumption of the implementation (line 7), taking into account component shutdown between different modes. Components can be shut down during the execution of a certain mode whenever no tasks belonging to that mode are mapped onto these components, i.e., the component is vacant (for instance, PE0 during execution of mode O_2 in Figure 5.5, page 107). Another important aspect is the reconfigurability of FPGAs which allows to exchange the implemented cores to suit the active mode. However, this reconfiguration during a mode change takes time, hence, a transition penalty is introduced if the maximal transition times are exceeded (line 8). Having determined the cores to be implemented (line 5), it is now also possible to schedule each mode of the application and to derive a feasible communication mapping (line 10). Since the modes are mutually exclusive, it is possible to employ a communication mapping and scheduling optimisation for a single mode system. The technique outlined in Section 4.2.2 is utilised for this purpose. If timing constraints are violated by the found schedule, a timing penalty is introduced (line 11). Furthermore, based on the communication mapping and scheduling, the dynamic power consumption of the application can be computed, taking into account DVS (line 12) if voltage-scalable components are present. Similarly to the shutdown of PEs, it is also possible to switch off a CL when no communications are mapped to that link (line 13), therefore, further reducing the static power consumption of the system. Based upon all estimated power consumptions and penalties, a fitness is calculated (line 14) as,

$$F_M = \bar{p} \cdot tp \cdot \left(1 + w_A \cdot \sum_{\pi \in \mathcal{P}_v} (a_\pi^U - a_\pi^{max})/(a_\pi^{max} \cdot 0.01)\right) \cdot \left(w_R \cdot \prod_{T \in \Theta_v} t_T/t_T^{max}\right)$$

(5.5)

where the average power dissipation $\bar{p}$ is given by Equation (5.1) and tp introduces a timing penalty if the schedule exceeds task deadlines or the repetition period. Further, an area penalty is applied for all PEs with area violation $\mathcal{P}_v$ by relating used area a_π^U and area constraint a_π^{max}. Similarly, a transition time penalty is applied for all transitions Θ_v that exceed their maximal transition time limit, i.e., transition time t_T exceeds the maximal allowed transition time t_T^{max}. Both area and transition penalty are weighted (w_A and w_R), which allows to adjust the aggressiveness of the penalty. Having assigned a fitness to all individuals of the population, they are ranked using linear scaling (line 15). A tournament selection scheme is used to pick individuals (line 16) for mating (line 17). The produced offsprings are inserted into the population (line 18). In order to improve the performance of the genetic algorithm, we apply four genetic mutation strategies that add problem specific knowledge into the optimization process (lines 19–22). This is achieved by introducing a small number of mutated individuals into the current population whenever the optimization process occurs to be trapped. These newly injected solution candidates provide the potential to turn into high quality solution by mating with other solution. These mutation strategies are introduced next.

Shutdown Improvement: To increase the chances of component shutdown, which leads to a reduction of static power consumption, the genetic task mapping algorithm employs a simple yet effective strategy during the optimisation. Out of the current population randomly picked individuals (probability 2% was found to lead to good results) are modified as follows. A single mode O_x and a non-essential PE π_a are selected. Non-essential PEs are considered to be PEs that implement task types that have alternative implementations on other PEs, hence, they are not fundamental for a feasible solution. Our goal is to switch off PE π_a during the execution of mode O_x. Therefore, all tasks of mode O_x which are mapped to π_a are randomly re-mapped to the remaining PEs ($\mathcal{P} \setminus \pi_a$), hence, PE π_a can be shut down during mode O_x. Of course, only feasible mappings are allowed, i.e., tasks are always mapped randomly to the PEs that are capable of executing this kind of task type. The pseudo code of this shutdown improvement strategy is show in Figure 5.9. A list of non-essential PE, i.e., PEs that have the potential to be switched off, is produced in line 01. The operational mode O_x is selected randomly, however, the greater the execution probability of a mode the higher are the chances to select this mode (line 02–08). The PE to be switched-off π_a is randomly selected in the lines 09–12. PEs accommodating less tasks than others are selected with a higher probability, since re-mapping only a small number of tasks is likely to influence the execution behaviour less drastically than the re-mapping of many tasks. After the operational mode O_x and the PE π_a have been selected, all tasks mapped to PE π_a in mode O_x are randomly re-mapped to the remaining PEs (line 13–14). The so mutated string is returned and inserted into the population pool. Due to the functional similarity of the four improvement strategies, only the pseudo code of the shutdown technique is given here (see Figure 5.9).

Area Improvement: To avoid convergence towards area infeasible solutions, a second strategy is employed. If only infeasible area mappings have been produced for a certain number of generations, the search is pushed away

---

Algorithm: *SHUTDOWN-IMPROVEMENT*

Input: - Randomly picked initial mapping string M (probability 2%)
        - OMSM (finite state machine + task graphs)
        - Technology Library, Allocated Architecture
Output: - For shutdown modified mapping string

```
01: NonEssPEs ← PossibleShutdownPEs()
02: Rand = RandomFloat(0, 1)
03: AccuProb = 0
04: O_x = 0
05: forall (Mode, ModeNumbers) {
06:     AccuProb += GetExecutionProb(Mode)
07:     if (RandDouble <= AccuProb) break
08:     else O_x++ }
09: Rand = RandomFloat(0,TasksNoInvSum)
10: forall(NoOfPE, NoAllocPEs) {
11:     if ((InvInterval0[NoOfPE] <= RandDouble) &&
            (InvInterval1[NoOfPE] > RandDouble)) {
12:         π_a = NoOfPE } }
13: forall (gene ≠ π_a) ∈ M
14:     gene = RandomFeasibleMapping(P \ π_a)
15: return M    // modified mapping
```

Figure 5.9. Pseudo Code: Mapping Modification towards component shutdown

from this region by randomly re-mapping hardware tasks onto software-programmable PEs.

Timing Improvement: In contrast to the area improvement strategy, if a certain amount of timing infeasible solutions have been produced, software tasks are randomly mapped to faster hardware implementations. Thereby, the chance to find timing feasible implementations is increased.

Transition Improvement: Cores implemented in FPGAs can be dynamically reconfigured. However, this involves a time overhead. If this overhead exceeds the imposed transition time limits, the mapping is infeasible. Hence, after generating for a certain number of generations solely solutions that violate the transition times, tasks are randomly re-mapped away from the FPGAs that cause the violations.

Although some of the produced genomes (strings) might be infeasible in terms of area and timing behaviour, all these strategies have been found to improve the search process significantly by introducing individuals that evolve into high quality solutions. For instance, running the synthesis process (on examples of moderate size) without the shutdown improvement strategy often results in implementations which do not exploit this energy reduction possibility.

5.5.3   Hardware Core Allocation

For tasks mapped to ASICs and FPGAs it is necessary to allocate hardware cores that are capable of executing the task types. This is a trivial job as long as only tasks of different types are mapped to the same hardware component, i.e., when a single core for each task needs to be allocated. Nevertheless, if tasks of the *same* type η are assigned to the *same* PE more than once, it is necessary to make a decision upon how many core of type η need to be implemented. This is important because hardware cores are able the execute tasks in parallel, i.e., the right quantitative choice of cores can efficiently help to exploit application parallelism, hence, improve the timing behaviour as well as energy dissipation. In the proposed co-synthesis, the following approach is employed during the schedule optimisation. Initially, each task type assigned to hardware is implemented only once, even if multiple tasks of this type are mapped onto the same PE. This ensures that all hardware tasks have at least one executable core implementation. If the hardware area constraints are not violated through the initial allocation, additional cores are implemented as follows. The tasks are analysed to identify possibly parallel executing task, taking into account task dependencies. These tasks are then ordered according to their mobility. Clearly, tasks with low mobility are more likely to improve the timing behaviour and therefore should be the preferred choice when implementing additional hardware cores. Accordingly, cores for tasks with low mobility are implemented as long as the area constraints of the hardware components are not violated. Note, this strategy potentially improves the energy dissipation, since it is probable to result in more slack time, which, in turn, can be exploited through DVS.

5.5.4   Dynamic Voltage Scaling for Multiple Parallel Executing Tasks

Dynamic voltage scaling is a powerful technique to reduce energy consumption by exploiting temporal performance requirements through dynamically adapting processing speed and supply voltage of PEs. An effective voltage scaling technique for this purpose has already been introduced in Chapter 3. Furthermore, the applicability of DVS to embedded distributed systems was demonstrated in [68, 100]. However, these works as well as the energy-gradient technique introduced in Chapter 3 concentrate on dynamically changing the performance of software PEs only, while parallel execution of tasks on hardware resources has been neglected. Nevertheless, in the context of energy-efficient multi-mode systems, where performance requirements of each op-

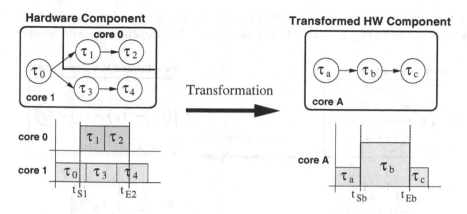

Figure 5.10. DVS Transformation for HW Cores

erational mode can vary significantly, DVS needs to be considered carefully. Consider, for instance, an inverse discrete cosine transformation (IDCT) algorithm implemented in fast hardware which is used during two modes: MP3 decoding and JPEG image decoding. Clearly, the JPEG decoder should restore images as quickly as possible, i.e., the IDCT hardware is required to execute at maximal supply voltage (equivalent to peak performance). On the other hand, the MP3 decoder works at a fixed repetition rate of $25ms$ for which the hardware implementation operates faster than necessary, i.e, the IDCT performance can be reduced such that this repetition rate is adequately met. By using DVS it is possible to adapt the execution speed to suit both needs and to reduce the energy consumption to a minimum.

This chapter considers that hardware components might employ DVS. However, due to the area and power overhead involved in additional DVS circuitry (DC/DC-converter [65, 109]) it is assumed that all cores allocated to the same hardware component are fed by a single voltage supply, i.e., dynamically scaling the supply voltage simultaneously affects the performance of all cores on that hardware component. To cope with this problem, the potentially parallel executing tasks on a single scalable hardware resource are transformed into an equivalent set of sequentially executing tasks, taking into account the dynamic power dissipation on each core. Note that this is done to calculate the scaled supply voltages only, i.e., this virtual transformation does not affect the real implementation. Figure 5.10 shows the transformation of five hardware tasks, executing on two cores (both cores are implemented within the same hardware component), to three sequential tasks on a single core. This sequential execution is equivalent to the behaviour of software tasks, hence, a voltage scaling technique for software processors can be applied. Nevertheless, it is important to consider task dependencies and task deadlines during this transformation, i.e., it might be necessary to further subdivide the transformed task graph, in

order to maintain a correct specification. For instance, consider original specification given in Figure 5.11. Here, task τ_3 has a data dependency with another

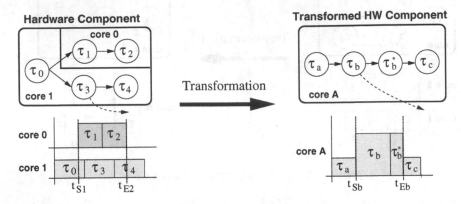

Figure 5.11. DVS Transformation for HW Cores considering inter-PE communication

task outside of the given component. Due to the data dependencies the transformation results in four tasks, instead of three (Figure 5.10). Thereby, task τ_b and
task τ_b^* can be scaled independent of each other. While task τ_b can influence
tasks on other PEs, task τ_b^* solely affects downstream tasks. Exactly as the
original specification. Task deadlines can be handled similarly.

The pseudo code description of this transformation is given in Figure 5.12.
The function takes as input the system task graph and a hardware processing
element that is DVS enabled. In the first two steps (lines 1 and 2) it generates
two priority queues of tasks mapped to the HW processing element, sorted in
decreasing order of start and end time, respectively. A sequential task graph and
a current power variable are initialised (lines 3 and 4). At the end of algorithm
run, the sequential task graph will hold the transformed tasks. The following
transformations are repeated until no tasks are left in the priority queue of task
sorted according to their end times (EndTaskQueue):

- If the next chronological event is due to a starting task (line 6), the current
  power dissipation is increased by the power dissipation of the starting task
  or tasks (line 7) and a new task is added to the sequential task graph by
  inheriting the current power dissipation and the start time (line 8). The
  starting tasks are remove from StartTaskQueue (line 9).

- If the next chronological event happens due to an end task (line 11), then
  the current power dissipation is decreased by the power of the ending tasks
  (line 12). According to whether the current power has reach zero or not, a
  task is added to the sequential task graph by inheriting the current power
  value and the outedges (if any) of the ending task (line 14), or the end time

---

Algorithm: *DVS-HARDWARE-TRANSFORMATION*

Input: - Task Graphs
- DVS-enabled Hardware Processing Element
Output:- Sequentialised tasks on the HW PE

```
01: StartTaskQueue = InitStartTaskQueue(Tasks)
02: EndTaskQueue = InitEndTaskQueue(Tasks)
03: SeqTaskGraph ← 0
04: Pw ← 0
05: while (EndTaskQueue ≠ 0) {
06:    if (TimeStartTask < TimeEndTask) {
07:        Pw += Pw(StartTasks)
08:        AddTask(SeqTaskGraph, Pw, StartTime)
09:        RemoveStartTasks(StartTaskQueue)
10:    }
11:    else if (TimeEndTask < TimeStartTask) {
12:        Pw -=Pw(EndTasks)
13:        if (Pw ≠ 0)
14:            AddTask(SeqTaskGraph, Pw, OutEdgesOffPE)
15:        else
16:            InheritEndTime(SeqTaskGraph->Last, OutEdgesOffPE)
17:        RemoveEndTasks(EndTaskQueue)
18:    }
19:    else if (TimeStartTask = TimeEndTask) {
20:        if (Core(StartTasks) = Core(EndTasks)) {
21:            if (OutEdgesOffPE(EndTasks) ≠ 0)
22:                AddTask(SeqTaskGraph, Pw, OutEdgesOffPE)
23:        }
24:        else {  // Core(StartTasks) ≠ Core(EndTasks) {
25:            Pw = Pw + Pw(StartTasks) - Pw(EndTask)
26:            AddTask(SeqTaskGraph, Pw, OutEdgesOffPE)
27:        }
28:        RemoveStartTasks(StartTaskQueue)
29:        RemoveEndTasks(EndTaskQueue)
30:    }
31: }
32: return (SeqTaskGraph)
```

Figure 5.12. Pseudo code: Task graph transformation for DVS-enabled hardware cores

and the outedges are inherited to the last task in the sequential task graph (line 16). The ending tasks are removed from `EndTaskQueue` (line 17).

- If the next chronological events occur to the simultaneously ending and starting task the following procedure is necessary (line 16). If both events appear on the same HW core (i.e., they subsequent tasks), and there are outedges emitting from the ending task, then a new task is added to the sequential task graph. However, the current power is maintained at the same level. Nevertheless, if the two events occur on different cores the current power is adjusted and an new task is introduced to the sequential task graph. Starting and ending tasks are removed StartTaskQueue and EndTaskQueue (lines 28 and 29).

After no tasks are left in the task queue, which is sorted in chronologically decreasing order of end times, the sequential task graph is returned (line 32).

5.6 Experimental Results: Multi-Mode

Based on techniques and algorithms presented in this chapter, the multi-mode synthesis approach has been implemented on a Pentium III/1.2GHz Linux PC. In order to evaluate its capability to produce high quality solutions in terms of energy consumption, timing behaviour, and hardware area requirements, a set of experiments has been carried out on 15 automatically generated multi-mode examples (mul1–mul15[1]) and one real-life benchmark example (smart-phone). The following two sections split these experiments into hypothetical and real-life examples. All reported results were obtained by running the optimisation processes 40 times and averaging the outcomes.

5.6.1 Hypothetical Examples

Each of the 15 generated examples (mul1–mul15) is specified by 3 to 5 operational modes, each consisting of 8 to 32 tasks (required execution cycles vary between 500–350000). The used target architectures contain 2 to 4 heterogeneous PEs (clock frequencies are given in the range from 25 to 50MHz), some of which are DVS enabled. These PEs are interconnected through 1 to 3 communication links. The active power consumption of programmable processors was randomly chosen between 5mW and 500mW, depending on the executed task. The power dissipation of hardware components are selected to be 1 to 2 orders of magnitude lower. Further, the static power dissipation was set to be 5 to 15% of the maximal active power. The execution probabilities of individual modes were randomly chosen and vary between 1% and 85%. Timing constraints have been assigned in the form of individual task deadlines as well as repetition periods to the modes (hyper-periods). The timing constraints were varied between 15ms and 500ms, such that schedulable implementations with up to 50% deadline slack could be found.

To illustrate the importance of taking mode execution probabilities into account during the synthesis process, an execution probability neglecting approach is compared with the proposed synthesis technique, which considers the mode probabilities. The first two sets of experiments demonstrate the energy savings achievable through the consideration of mode execution probabilities, either with or without the exploration of DVS. The third set examines the influence of the actual activation profile on the energy savings.

Comparisons excluding Dynamic Voltage Scaling

To highlight the influence of mode execution probabilities on the achievable energy savings, consider Table 5.2 which shows the multi-mode co-synthesis results for the 15 automatically generated benchmarks, distinguishing between an execution probability neglect approach ("w/o probab.") and the introduced approach ("with probabilities"). The first three columns give the benchmark

Example	Hyper-period $(10^{-2}s)$	Mode Execution Probabilities	w/o probab.		with probabilities		
			Aver. power (mW)	CPU time (s)	Aver. power (mW)	CPU time (s)	Red. $(\%)$
mul1	7/6/9/2	5:10:75:10	8.131	20.7	7.529	24.7	7.29
mul2	5/7/4/8	10:5:80:5	3.404	15.5	2.771	18.2	18.61
mul3	2/2.4/6/4/3	7:3:80:5:5	10.923	23.4	10.430	23.0	4.17
mul4	6/3/7/4/5	1:4:5:40:50	7.975	21.0	6.726	25.2	15.50
mul5	2/6/6/4	7:13:35:45	5.186	18.4	4.668	22.1	10.01
mul6	6.5/4/4/10	15:10:10:65	1.677	20.6	1.301	19.9	22.46
mul7	20/16/19/10	5:5:5:85	3.306	11.6	1.250	21.4	62.18
mul8	40/7/4/8	75:5:15:5	1.565	32.1	1.329	28.0	15.06
mul9	4/4/10/4	7:3:80:10	3.081	6.0	1.901	5.8	38.28
mul10	50/7/50/8/7	45:5:40:5:5	1.105	28.3	0.941	32.1	14.83
mul11	10/12/20	80:10:10	2.199	9.3	1.304	16.6	40.70
mul12	8/8/9/15	15:10:50:25	7.006	25.4	5.975	34.2	14.69
mul13	8/6/10	5:15:80	4.090	15.8	2.816	15.8	31.04
mul14	6/3/6/10/19	5:5:10:10:70	8.195	28.6	6.466	33.0	21.13
mul15	22/15/6/25/20	5:10:75:7:3	2.188	41.5	1.222	55.4	44.16

Table 5.2. Considering mode execution probabilities (excluding DVS)

names, the hyper-period (repetition period) of each mode, and the mode execution probabilities. The fourth and the fifth column present the dissipated average power and optimization time for the execution probability neglecting synthesis approach. Note that the execution probabilities are neglected during the synthesis only, while the computed power dissipations at the end of the synthesis incorporate the execution probabilities, in order to ensure a meaningful comparison. The sixth and the seventh column show the same for the proposed approach, which considers the execution probabilities throughout the synthesis process. Take, for instance, example mul6. When ignoring the execution probabilities during the optimization, an average power dissipation of $1.677mW$ is achieved. However, optimizing the same benchmark example under the consideration that modes execute with *uneven* probabilities (e.g., 15:10:10:65 — i.e., mode 1 is active for 15%, mode 2 is active of 10%, and so on), the average power can be reduced by an appropriate task mapping and core allocation to $1.301mW$. This is a significant reduction of 22.46%. Furthermore, it can be observed that the proposed technique was able to reduce the energy consumption of all examples with up to 62.18% (mul7). Note that these reductions are achieved without any modification of the underlying hardware architectures, i.e., the system costs are not increased. It is also important to note that the achieved energy reductions are solely introduced by taking the mode execution probabilities into account during the co-synthesis process, i.e., both compared approaches allow the same resource sharing and rely on the same scheduling technique. When comparing the optimization times for both approaches, it can be observed that the proposed technique shows a slightly increased CPU time for most examples, which is mainly due to the more complex design space structure.

Comparisons including Dynamic Voltage Scaling

The next experiments were conducted to see how the proposed technique compares to DVS and if further savings can be achieved by taking the mode probabilities and DVS simultaneously into account. Table 5.3 reports on the findings. The DVS technique that was used here is based on PV-DVS [133], which has been extended to enable the consideration of DVS not only for software processors, but also for parallel executing cores on hardware PEs (see Section 5.5.4). As in the first experiments, two approaches are compared here. The first approach disregards the mode execution probabilities during optimization, while the second takes them into account throughout the co-synthesis. Similar to Table 5.2, the second and the third column of Table 5.3 show the results without consideration of execution probabilities, whilst the fourth and the fifth column present the results achieved by taking execution probabilities into account. Let us consider again benchmark mul6. Although the execution probabilities are neglected in the fourth column, a reduced average power consumption $(0.689mW)$ can be observed, when compared to the results given in Table 5.2. This clearly demonstrates the high energy reduction capabilities of DVS. Nevertheless, it is possible to further minimize the power consumption to $0.465mW$ by considering the execution probabilities together with DVS. This is an improvement of 32.53%, solely due to the synthesis for the particular execution probabilities. For all other benchmarks savings of up to 64.02% (mul7) were achieved. Due to the computation of scaled supply voltages and the influence of scheduling on the energy consumption, the optimization times are higher when DVS is considered.

Example	Hyper-period $(10^{-2}s)$	Mode Execution Probabilities	w/o probab.		with probabilities		
			Aver. power (mW)	CPU time (s)	Aver. power (mW)	CPU time (s)	Red. $(\%)$
mul1	7/6/9/2	5:10:75:10	4.271	526.6	3.964	768.6	10.92
mul2	5/7/4/8	10:5:80:5	1.568	860.4	1.273	687.4	18.82
mul3	2/2.4/6/4/3	7:3:80:5:5	4.012	1053.5	3.344	1192.2	16.66
mul4	6/3/7/4/5	1:4:5:40:50	2.914	1135.2	2.320	1125.4	20.39
mul5	2/6/6/4	7:13:35:45	1.394	967.7	1.315	932.1	5.68
mul6	6.5/4/4/10	15:10:10:65	0.689	472.9	0.465	593.7	32.53
mul7	20/16/19/10	5:5:5:85	1.331	540.3	0.479	820.7	64.02
mul8	40/7/4/8	75:5:15:5	0.564	1262.1	0.436	1412.0	22.64
mul9	4/4/10/4	7:3:80:10	0.942	161.2	0.648	177.1	34.66
mul10	50/7/50/8/7	45:5:40:5:5	0.480	1456.3	0.394	1361.9	17.88
mul11	10/12/20	80:10:10	0.396	318.1	0.255	403.2	35.53
mul12	8/8/9/15	15:10:50:25	2.857	1384.7	2.460	1450.7	13.91
mul13	8/6/10	5:15:80	1.185	498.3	0.953	576.6	19.56
mul14	6/3/6/10/19	5:5:10:10:70	2.320	1512.3	1.797	1556.4	22.55
mul15	22/15/6/25/20	5:10:75:7:3	0.801	1316.7	0.324	1836.3	59.56

Table 5.3. Considering mode execution probabilities (including DVS)

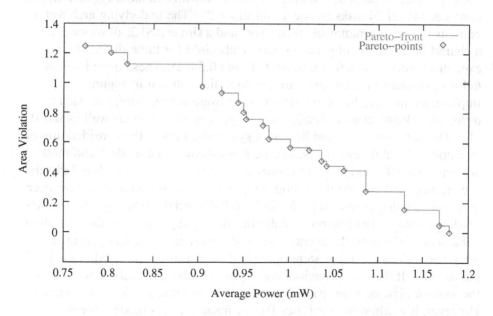

Figure 5.13. Pareto optimal solution space achieved through a single optimisation run of mul15 (without DVS), revealing the solution trade-offs between energy dissipation and area usage

Figure 5.13 shows the Pareto outcome of the typical co-synthesis run. Each diamond in this graph indicates a possible implementation of benchmark mul15, however, each with a different trade-off between energy dissipation of area usage. The rightmost solution respects all imposed system constraints including area (area violation is 0). This solution dissipates an energy of $1.179mW$. Nevertheless, the graph indicates that further energy savings can be made by mapping more functions towards hardware. Of course, increasing the used area corresponds to more costly design. For instance take the leftmost solution, which dissipated $0.773mW$. To achieve this reduced energy dissipation it is necessary to increase the available hardware area by 124%, i.e., the hardware requirements are more than doubled.

Influence of Real Activation Probabilities

The next experiment is conducted to highlight the influence of the user behaviour on the energy efficiency of a system that has been synthesised under the consideration of certain mode execution probabilities. Certainly, the mode execution probabilities, which are used during the synthesis represent an "imaginative" user and the activation probabilities of a *real* user will differ from those. Accordingly, the following experiment tries to answer the question how the energy efficiency is affected by different activation profiles during application run-time. For experimental purpose a simple specification with two modes is used which contains 14 and 24 tasks (mode 1 and mode 2). The underlying architecture consists of two programmable processors and a single ASIC, all connected via a shared bus. This configuration was synthesised for three different pairs of execution probabilities (0.1:0.9, 0.9:0.1, and 0.5:0.5). These three implementations possibilities correspond to the three lines shown in Figure 5.14. All implementations are based on the same hardware architecture; yet, each has a different task and communication mapping, core allocation, as well as schedule. The first solution (solid line) was synthesised under the consideration of execution probabilities 0.1:0.9, that is, it is assumed that mode 1 and mode 2 are active for 10% and 90% of the operational time, respectively. Similarly, the second (dashed) and the third (dotted) line represent solutions that have been synthesised using execution probabilities 0.9:0.1 and 0.5:0.5, respectively. According to the real execution probabilities during run-time, i.e., the activation behaviour of the user, the average power dissipations of the implemented systems vary. Consider the system optimised for execution probabilities 0.1:0.9 (solid line). If the user behaviour corresponds to these probabilities (User A), the system dissipates an average power of approximately $5.5mW$ (point I). However, if a different user (User B), for instance, uses mode 1 for 90% and mode 2 for 10% of the time (0.9:0.1), the system will dissipate approximately $12.5mW$ (point II). Nevertheless, if the system would be optimised for this activation profile (0.9:0.1), as indicated by the dashed line in Figure 5.14, a lower power dissipation of around $2.6mW$ (point III) can be achieved. Similarly, if the system is optimised for execution probabilities 0.9:0.1 (dashed line) and the

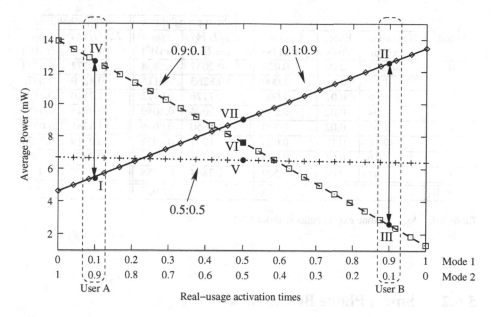

Figure 5.14. A system specification consisting of two operational modes optimized for three different execution probabilities
(solid line–0.1:0.9, dashed–0.9:0.1, dotted–0.5:0.5)

user runs the application 10% in mode 1 and 90% in mode 2 (User A), then a power dissipation of $12.5mW$ (point IV) is given. While an optimisation towards this usage profile can achieve a system implementation which dissipates only $5.5mW$ (point I), i.e., extending the battery-lifetime by a factor of 2.83 times. The dotted plot in Figure 5.14 represents the solution when the execution probabilities are neglected during the optimisation, that is, the execution probabilities are considered to be equal for both modes. Of course, if the modes 1 and 2 are active for equal amounts of time, this solution achieves a lower power dissipation ($6.5mW$, point V) than the systems optimised for execution probabilities 0.1:0.9 ($9mW$, point VI) and 0.9:0.1 ($7.6mW$, point VII). The figure reveals that the design for 0.1:0.9 (solid line) achieves the lowest power dissipation when the user complies to an activation profile between 0:1 and 0.21:0.79. While the designs for 0.5:0.5 (dotted line) and for 0.9:0.1 (dashed line) lead to the lowest energy dissipation in the ranges from 0.21:0.79 to 0.57:0.43 and 0.57:0.43 to 1:0, respectively. In summary, Figure 5.14 clearly shows that the execution probabilities substantially influence the energy dissipation of the system. Certainly, the system should be optimised as close as possible towards the real behaviour to achieve low energy consumptions, which, in turn, result in longer battery-lifetimes.

Mode	No. tasks/ comm.	Exec. Prob.	Hyper- period (s)	without probabilities		with probabilities	
				E_O/HP_O (mJ)	$\bar{p}_O$ (mW)	E_O/HP_O (mJ)	$\bar{p}_O$ (mW)
0	88/137	0.09	0.020	0.2637	1.1868	0.1272	0.5723
1	12/0	0.74	1.000	0.8263	0.6115	0.8210	0.6075
2	12/0	0.01	1.000	1.7176	0.0172	1.7110	0.0171
3	5/4	0.02	0.250	1.3004	0.1040	0.9545	0.0764
4	12/5	0.02	0.500	1.6650	0.0666	1.3761	0.0550
5	17/16	0.10	0.025	0.1231	0.4922	0.0719	0.2874
6	17/16	0.01	0.025	0.2203	0.0881	0.3971	0.1588
7	12/5	0.01	0.500	1.7884	0.0358	1.3245	0.0265
Overall					2.6022		1.8011

Table 5.4.   Smart phone experiments without DVS

5.6.2   Smart Phone Benchmark

To further validate the co-synthesis technique in terms of real-world applicability, the introduced approach was applied to a smart-phone example. This benchmark is based on three publicly available applications: a GSM codec [4], a JPEG codec [5], and an MP3 decoder [72]. Accordingly, the smart-phone offers three different services to the user, namely, a GSM cellular phone, a digital camera, and an MP3-player. Of course, the used applications do not specify the whole smart-phone device, however, a major digital part of it. The specification for this example, given as operational mode state machine (OMSM), has already been introduced in Figure 5.1. For each of the eight operational modes, the corresponding task graphs have been extracted from the above given references. The individual applications have been software profiled to gather the necessary execution characteristics of each task. This was carried out by compiling profile information into the application [1, 3] and running the produced software on real-life input streams. On the other hand, the hardware estimations are not based on direct measurements, but have been based on typical values, such that hardware tasks typically executed 1 to 2 orders of magnitude faster and dissipated 1 to 2 orders of magnitude less power than their software counterparts [35]. Depending on the operational mode, the number of tasks and communications varies between 5–88 nodes and 0–137 edges, respectively. The hardware architecture of the embedded system within the smart phone consists of one DVS-enabled processor (execution properties are based on values given for the ARM8 developed in [34]) and two ASICs. These components are connected via a single bus. Tables 5.4 and 5.5 give the results of the conducted experiments, distinguishing between optimizations without and with the consideration of DVS.

Mode	No. tasks/ comm.	Exec. Prob.	Hyper- period (s)	without probabilities		with probabilities	
				E_O/HP_O (mJ)	$\bar{p}_O$ (mW)	E_O/HP_O (mJ)	$\bar{p}_O$ (mW)
0	88/137	0.09	0.020	0.0746	0.3355	0.0786	0.3539
1	12/0	0.74	1.000	0.8190	0.6061	0.0180	0.0133
2	12/0	0.01	1.000	0.0280	0.0003	0.8110	0.0081
3	5/4	0.02	0.250	0.3355	0.0268	0.3545	0.0284
4	12/5	0.02	0.500	0.3556	0.0142	0.8412	0.0336
5	17/16	0.10	0.025	0.0513	0.2052	0.0813	0.3250
6	17/16	0.01	0.025	0.0492	0.0197	0.1975	0.0791
7	12/5	0.01	0.500	0.4917	0.0098	0.8671	0.0173
Overall					1.2176		0.8587

Table 5.5. Smart phone experiments with DVS

Similar to the previous experiments, approaches which neglect the execution probabilities are compared with the introduced co-synthesis technique that considers the uneven activation times of different modes. Table 5.4 shows this comparison for a fixed voltage system, i.e., no DVS is applied. The table provides information regarding all 8 modes of the smart phone. This mode information includes benchmark properties such as complexity, execution probability, and hyper-period. Furthermore, the table gives the achieve energy dissipation for the mode hyper-period and average power consumption of each mode. The average power consumption can be calculated from the energy values by dividing the energy by the hyper-period and multiplying the result with the execution probability. Synthesizing the system without consideration of execution probabilities results in an overall average power consumption of $2.6022mW$, when running the system after the synthesis according to the activation profile. Nevertheless, taking into account the mode usage profile during the co-synthesis this can be reduced by 30.76% to $1.8011mW$. Please note that the given overall average power consumption is calculated based on Equations (5.1)–(5.4); hence, these values are directly proportional to the battery-life time. The saving is achieved without the modification of the allocated hardware architecture, therefore, the system cost is the same for both solutions.

Also DVS has been applied to this benchmark, considering that the GPP of the given architecture supports DVS functionality. The results are shown in Table 5.5. It can be observed that the overall average power consumption of the smart phone drops to $1.2176mW$, even when neglecting mode execution probabilities. However, the combination of applying DVS and taking execution probabilities into account results in the lowest power consumption of $0.8587mW$, a 29.5% reduction, when compared to the activation profile neglecting approach. That is, solely by considering the activation profile during the synthesis, the battery-life time could be extended by one third, even when using a system that employs DVS components. Overall, the average power is decreased from $2.602mW$ to $0.859mW$, which represents a significant reduction of nearly 67%. Regarding the required co-synthesis times, the four implementations could be found in $80.1s$ (without probabilities and DVS) to $4344.8s$ (with probabilities and DVS). Clearly, considering DVS requires longer optimization times due to the voltage scaling problem that needs to be solved repetitively within the innermost optimization loop of the co-synthesis algorithm. For instance, the optimization for DVS increases the run-time from $80.1s$ to $3754.1s$ for the case without consideration of execution probabilities, and from $96.9s$ to $4344.8s$ when execution probabilities are taken into account. On the other hand, the consideration of mode execution probabilities increases the optimization time only moderately form $80.1s$ to $96.9s$ in the case of no DVS, and from $3754.5s$ to $4344.8s$ if the probabilities are considered.

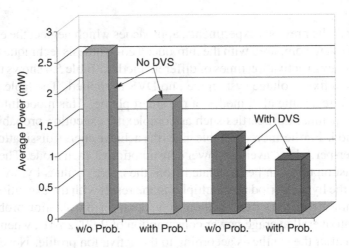

Figure 5.15.   Energy dissipation of the Smart phone using different optimisation strategies

5.7   Concluding Remarks

This chapter has introduced new techniques and algorithms for the energy minimisation of multi-mode embedded systems. An abstract specification model called operational mode state machine has been proposed. This model allows for the concurrent specification of mode interaction (top-level finite state machine) as well as mode functionality (task graph). The advantage of such a representation is the capability to express the complete functionality of the system within a single model, containing both control and data flow.

The presented co-synthesis technique not only optimises mapping and scheduling towards hardware cost and timing behaviour, but also aims at the reduction of power consumption at the same time. A key contribution in this chapter has been the development of an effective mapping strategy that considers uneven mode execution probabilities as well as important power reduction aspects, such as multiple task implementations and core allocation. For this purpose a GA based mapping approach has been proposed along with four improvement strategies to effectively handle the optimisation of component shutdown, transition time, area usage, and timing behaviour. These improvement strategies guide the mapping optimisation of multi-mode specifications towards high quality solutions in terms of power consumption, timing feasibility, and area usage. Furthermore, these strategies help to reduce the optimisation times needed by conventional genetic algorithms, since they use computational inexpensive constructive heuristic in a local search fashion. A newly introduced transformational-based algorithm for DVS-enabled hardware components, which is capable of performing parallel task execution, allows to easily leverage the efficiency of existing voltage scaling algorithms. This algorithm transforms a set of potentially parallel executing tasks on a single HW component into a set of sequential executing tasks, taking into account imposed deadlines and inter-PE communications.

The proposed techniques and algorithms have been validated through extensive experiments including a smart phone real-life example. These experiments have demonstrated that taking into account mode execution probabilities throughout the system synthesis leads to substantial energy savings compared to conventional approaches which neglect this issue. Furthermore, DVS has been considered in the context of multi-mode embedded systems and it was shown that considerably high energy reductions can be achieved by combining both the consideration of execution probabilities and dynamic voltage scaling.

Notes

1 These examples were generated with the publicly available tool TGFF [48].

Chapter 6

DYNAMIC VOLTAGE SCALING FOR CONTROL FLOW-INTENSIVE APPLICATIONS

BY DONG WU, BASHIR M. AL-HASHIMI, AND PETRU ELES

Up to now, the applications that have been addressed in this book are considered to be data-flow dominated. Such applications (e.g. voice encoding and decoding, image and video processing) can be accurately modelled using a task graph representation with the opportunity to express limited control-flow inside each task specification. However, for applications with an extensive, global control-flow, models such as the conditional task graphs (CTG) [1] [51, 159] represent an adequate choice, since they allow a more precise and thorough capturing of this type of system behaviour. One major problem in the presence of control flow is the uncertainty regarding which tasks on different control-paths will be scheduled at run-time, i.e., scheduling decisions in terms of voltage settings and task executions have to be taken during the execution of the application.

In this section, we will concentrate on dynamic voltage scaling and scheduling approaches for applications that are captured through conditional task graphs. The chapter is organised as follows: Preliminaries regarding the CTG model and current scheduling techniques for applications modelled as CTGs are introduced in Sections 6.1 and 6.2. The problems involved in dynamic voltage scaling of CTG specifications are outlined in Section 6.3. This is followed by a dynamic voltage scaling approach for applications modelled by CTGs in Section 6.4. Conclusions are drawn in Section 6.5.

6.1 The Conditional Task Graph Model

In this chapter, we consider that an application has been specified as a directed, acyclic graph $G(\mathcal{T}, \mathcal{E}_U, \mathcal{E}_C)$ called Conditional Task Graph (CTG) [50, 51]. An example CTG is shown in Figure 6.1(a). In this model, each vertex represents a tasks $\tau_i \in \mathcal{T}$. These tasks are connected through two different types of edges, denoted as unconditional edges $\mathcal{E}_U$ and conditional edges $\mathcal{E}_C$. The

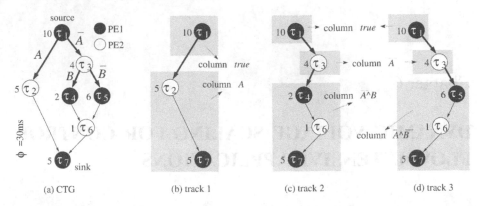

Figure 6.1.   Conditional Task Graph and its Tracks

set of unconditional edges $\mathcal{E}_U$ and the set of conditional edges $\mathcal{E}_C$ are disjoint and form together the set of all edges $\mathcal{E}$, i.e., $\mathcal{E}_U \cap \mathcal{E}_C = \varnothing$ and $\mathcal{E} = \mathcal{E}_U \cup \mathcal{E}_C$. Similar to the task graph model introduced in Section 1.2.1, an edge $\gamma_{ij} \in \mathcal{E}$ indicates that the output of task τ_i is required as input to task τ_j. The main difference between the conditional task graph model used in this chapter and the previous task graph model is the additional existence of *conditional edges*, indicated with thick lines in Figure 6.1. Each conditional edge $\gamma_{ij} \in \mathcal{E}_C$ is associated with a condition value. In order to perform a transition on such an edge, the condition value has to be true. For instance, depending on the outcome of task τ_1, the condition value A or $\bar{A}$ is produced. Accordingly, either task τ_2 (if A is true) or task τ_3 (if $\bar{A}$ is true) is executed next. Nodes with outgoing conditional edges are denoted as *disjunction nodes* (τ_1 and τ_3 in Figure 6.1(a)), while alternative paths starting from the same disjunction node meet in a *conjunction node* (τ_6 and τ_7 in Figure 6.1(a)). A conjunction task can start its execution after the input of one of its alternative paths has arrived, as opposed to all other tasks which can be activated only after all their inputs have been received. Depending on the condition values, there exist different tracks through the CTG that may be followed during the execution. Clearly, as the condition values are only produced during run-time of the application, the actual tracks that are followed are unknown before execution. Nevertheless, all possible tracks can be identified before run-time. For instance, the Figures 6.1(b)–(d) show the three possible tracks of the CTG given in Figure 6.1(a). The execution delay of one instantiation of the CTG is given as the difference between the time when the sink node terminates execution and the start time of the source task. For instance, consider the CTG in Figure 6.1(a) and its three execution tracks shown in Figures 6.1(b)–(d). The numbers beneath each node indicate the execution time of that particular task (assuming a given mapping). In addition, the CTG is associated with a timing constraint Φ (repetition period) which represents the

	true	A	$\bar{A}$	$A \wedge B$	$\bar{A} \wedge \bar{B}$
τ_1	0,10				
τ_2		10,15			
τ_3			10,14		
τ_4				14,16	
τ_5					14,20
τ_6				16,17	20,21
τ_7		15,20		17,22	21,26

Table 6.1. Example Schedule Table for the CTG of Figure 6.1(a)

deadline of the sink node. For simplicity, we ignore here the additional delay imposed by communications between tasks.

A suitable application for a CTG specification could be, for example, an MPEG video decoder which reconstructs a video from a stream of "intra" pictures (I-frame), "predictive" pictures (P-frame), and "bidirectionally-predictive" pictures (B-frame) [149, 151]. While I-frames are coded without reference to other pictures (using similar techniques that are used for the compression of still images), P-frames and B-frames make use of the motion compensation technique to achieve a higher compression ratio. Accordingly, the CTG specification for such a decoder should use three alternative tracks, each responsible for the decoding of a certain frame type. The deadline of the CTG is set corresponding to the required frame rate. We refer the interested reader to [50, 51], where further details regarding the CTG model can be found.

6.2   Schedule Table for CTGs

During run-time, at each activation, a subset of tasks of the CTG is executed depending on the condition values that determine the actual track through the CTG. A scheduling algorithm for mapped conditional task graphs was introduced in [50]. The main aim of this technique is the minimisation of the worst case execution delay. The algorithm generates a *schedule table*. The produced schedule table contains task activation times for all nodes in the CTG, depending on the possible condition values. An example schedule table is given in Table 6.1, which is based on the execution times and mappings given in Figure 6.1(a). Each row in the schedule table corresponds to one task and represents the start time and end time for that task. Further, each column in the table corresponds to a logical expression which is constructed as a conjunction of condition values. Accordingly, the table holds the schedules for all possible tracks depending on the condition values. The execution tracks given in Figures 6.1(b)–(d) are represented in the table as follows: The schedule of track 1 is given in the columns *true* and A, i.e., task τ_1 runs from 0 to $10ms$, while tasks

τ_2 and τ_7 are active (if A is true) from 10–$15ms$ and 15–$20ms$, respectively. Similarly, the schedule of track 2 is captured in columns $true$, $\bar{A}$, and $\bar{A} \wedge B$. And finally, track 3 corresponds to columns $true$, $\bar{A}$, and $\bar{A} \wedge \bar{B}$.

The schedule table stores the quasi-static schedule of the system that is specified by a CTG for a given mapping of tasks. The off-line computed schedules, which are captured within the schedule table, are activated according to the condition values that are produced during run-time of the application. The real-time kernels running on each processing element will take the decisions regarding task activation depending on the condition values.

It should be noted that the scheduling is performed statically and not at run-time (on-line). The scheduling algorithm traverses the CTG, analysing the possible alternative tracks and considering for each track only the tasks that are activated for the respective condition values. Using a depth-first search, the algorithm generates the schedule table by proceeding along a binary decision tree that corresponds to alternative tracks [50]. Additionally, if dynamic voltage scaling is considered in the context of CTGs, it is necessary to associate a supply voltage with each task execution that determines at which speed and energy cost the specific task is executed. This is necessary in order to reduce the energy consumption to a minimum, while meeting imposed task deadlines. A suitable voltage scaling technique is introduced in the next section.

6.3   Dynamic Voltage Scaling for CTGs

The problems of dynamic voltage scaling in the context of CTGs are illustrated next, using the CTG shown in Figure 6.1(a). For the sake of this example, consider that the deadline of the CTG is set to $30ms$. The three possible schedules are depicted in Figure 6.2 as Gantt-diagrams for both processors of the embedded system. These three schedules correspond to the three possible tracks given in Figures 6.1(b)–(d). The schedules have been produced with the aim to reduce the worst case delays as much as possible (using the algorithm presented in [50]). As we can observe from the Figures 6.2(a)–(c), the amount of slack time that is exposed in the schedules varies with each track. For instance, track 1 offers $10ms$ of slack, while track 2 and track 3 reveal $8ms$ and $4ms$ of slack. Clearly, the amount of available slack varies during run-time depending on the condition values. In addition to the timing information, the Figures 6.2(a)–(c) give the energy consumption of each track, under the assumption that both processing elements have an active power dissipation of $5W$ at the nominal voltage of $3.3V$. Take for example track 1. Its energy dissipation is given by $5W \cdot (10ms + 5ms + 5ms) = 100mJ$. One possible way to exploit the slack that is exposed in the schedules of Figures 6.2(a)–(c), is to stretch the execution as much as possible, such that they fit the imposed deadline. The required *scaling factor* is calculated as the ratio between the

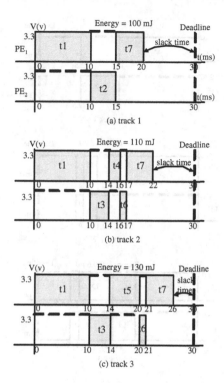

Figure 6.2. Schedules of the CTG of Figure 6.1(a) (in this figure t_i corresponds to tasks τ_i)

deadline and the length of the schedule. As an example, the scaling factor of track 1 is given by $30ms/20ms = 1.5$, that is, each task execution time can be prolonged by 1.5 without violating the specified deadline. In a similar fashion, the scaling factors for tracks 2 and 3 can be calculated as $30ms/22ms = 1.364$ and $30ms/26ms = 1.154$. The scaled schedule for each track is given in Figures 6.3(a)–(c). The schedules give the energy consumption of each track, assuming that the PEs' nominal supply voltage is $3.3V$. In Figure 6.3(a), for example, the extension from $20ms$ to $30ms$ allows us to decrease the supply voltages of both processing elements to $2.62V$ (in accordance to Equation (3.2) with d^* given by the scaling factor and a threshold voltage of $0.8V$). Thereby, reducing the energy dissipation from $100mJ$ to $62.9mJ$, based on:

$$E(V_{dd}) = E(V_{max}) \cdot \frac{V_{dd}^2}{V_{max}^2} \tag{6.1}$$

which can be derived from Equation (2.6). Coming back to the schedules given in Figure 6.3, we can observe that the execution time of task τ_1 differs on all three tracks, since the different amounts of slack time are exploited through different voltage settings (processors performances), that is, one and the same

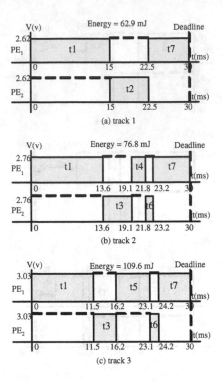

Figure 6.3.   Schedules scaled for energy minimisation

task is executed at a different voltage depending on the executed track. On track 1 it runs from 0 to $15ms$, while on tracks 2 and 3 the execution runs from 0 to $13.6ms$ and from 0 to $11.5ms$, respectively. Clearly, these execution times depend on the particular track and, by this, on the condition values which are not known in advance. Thereby, if the supply voltages and implicitly the execution times of tasks are not chosen properly, the system's timing constraints might be violated. To outline this in more detail, consider Figure 6.4. If, for instance,

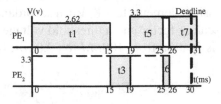

Figure 6.4.   Improper scaling with violated timing constraint

task τ_1 is executed from 0 to $15ms$ using the voltage setting for track 1 and the condition values turn out to be $\bar{A} \wedge \bar{B}$ at run-time (that is, track 3), the imposed deadline of $30ms$ will be missed, even when running the remaining tasks at maximum supply voltage. Thus, in order to exploit as much slack time as possible and, at the same time, meet the imposed timing constraints of the system, the worst case slack time should be identified dynamically. The worst case slack time is the maximum slack that can be distributed to a task such that no timing constraints are violated, no matter what schedule decisions are taken later (depending on the condition values). The goal of the DVS technique for CTGs introduced in this chapter, is the voltage selection such that, under any possible set of condition values, system deadlines are met and the energy consumption is minimised.

6.4 Voltage Scaling Technique for CTGs

In this section, we introduce a voltage scaling technique for applications that are specified as conditional task graphs (CTGs). The main principle behind the introduced approach is the identification of the available worst case slack, taking into account the conditional behaviour of the CTGs. This is achieved by dynamically identifying the worst case track, calculating a suitable scaling factor (i.e., the ratio between the deadline and the total length of the schedule), and adjusting the schedule table after a disjunction node (a node producing a condition value) has been scheduled. The initial schedule table, which is the input to the DVS technique introduced here, is produced using the approach presented in [50]. While the approach in [50] minimises the worst case execution delay, the introduced DVS technique produces a modified schedule table that indicates scaled voltages and activation times such that deadlines are fulfilled and the energy savings are optimised.

The introduced voltage scaling strategy is based on the idea to leverage the information concerning condition values that is available at a certain moment, in order to apply the largest possible scaling factor while still guaranteeing to meet the system deadlines. Such information regarding the conditional values becomes available whenever the execution of a disjunction node ends. Consequently, early in the scheduling process, more conservative scaling factors are applied, while with every available condition value, refined scaling factors can be calculated. For instance, if a disjunction node has been scheduled and for a certain track a longer slack than the one previously used can be identified, then this new slack can be taken to recalculate the applicable scaling factor. Of course, this increased scaling factor will further minimise the energy consumption compared to the previous (more conservative) scaling factor. According to this observation, the schedule of a CTG is divided into several *scaling regions* corresponding to the disjunction nodes. Each of the scaling regions is scaled with a certain, suitable scaling factor that will guarantee the fulfilment of timing constraints as well as the minimisation of the energy consumption. Examining Table 6.1, we can see that the schedules of the tasks in each of the columns

	true	A	Ā
τ_1	0,2		
τ_2	2,6		
τ_3	2,4		
τ_4		4,6	
τ_5			4,8
τ_6		6,8	8,10
τ_7		8,10	10,12

Table 6.2. Schedule Table for the CTG of Figure 6.5

correspond to such scaling regions. Please note, however, that columns in the initial schedule tables do not necessarily correspond to scaling regions directly. Consider the following situation: The CTG shown in Figure 6.5 is executed with condition value A and has to fulfil a timing constraint of $15ms$. According to

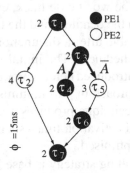

Figure 6.5. CTG with one disjunction node

schedule table 6.2, the resulting schedule is depict in Figure 6.6(a). As we can see from the schedule, task τ_2 is executed in parallel with disjunction task τ_3, which finishes execution before task τ_2, i.e., a condition value becomes available during the run-time of task τ_2. In our case, task τ_3 produces condition value A; thereby, selecting the corresponding track on which we can find task τ_4. Now that the track and the available worst case slack are given, it is possible to recalculate the scaling factor in order to make use of the additional slack. The scaled schedule and the corresponding scaled schedule table are given in Figure 6.6(b) and Table 6.3, respectively. As we can observe from the table, task τ_2 belongs to three scaling regions, one before the end of disjunction task τ_3 and two after.

In order to apply DVS without the violation of deadlines, it is necessary to identify scaling regions delimited by the end times of disjunction nodes and to scale the schedules of the tasks in each region after determining the available

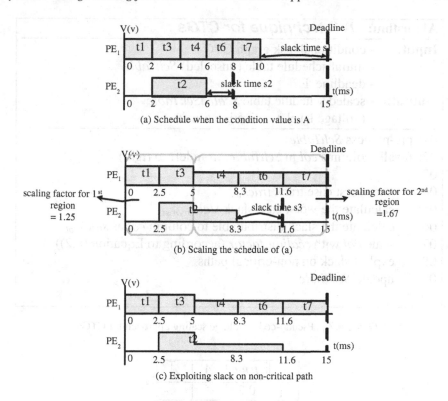

Figure 6.6.

	true	*A*	*Ā*
τ_1	0,2.5		
τ_2	2.5,5	5,8.3	5,7.5
τ_3	2.5,5		
τ_4		5,8.3	
τ_5			5,10
τ_6		8.3,11.6	10,12.5
τ_7		11.6,15	12.5,15

Table 6.3. Scaled Schedule Table for the CTG of Figure 6.5

slack and the appropriate scaling factor. A drawback of this scaling technique is that the tasks on the non-critical paths cannot take advantage of all available slack, as shown, for example, in Figure 6.6(b). Since task τ_2 does not lie on the critical path (which is given by τ_1, τ_3, τ_4, τ_6, and τ_7), it is not scaled to its full potential when using the scaling factor calculated for the worst case execution. Thereby, task τ_2 finishes execution at $8.3ms$, although its execution

Algorithm: **DVS technique for CTGs**
Input: - conditional task graph CTG
- initial schedule table (unscaled) *SchTable*
- deadline Φ
Output: - scaled schedule table *ScaledSchTable*
(voltage level + activation times)
01: pre-process *SchTable*
02: **forall** (columns *col* in *SchTable*, from left to right)
03: {
04:     find worst case track *track$_{worst}$*
05:     calculate the worst case slack *slack$_{worst}$*
06:     calculate the slack distributable to column *col* as *slack$_{col}$*
07:     scale *col* with *scaling_factor* (according to Equation (6.2))
08:     exploit slack on non-critical paths
09:     update *SchTable*
10: }

Figure 6.7.   Pseudo-code: Voltage scaling approach for CTGs

	true	A	$\bar{A}$
τ_1	0,2		
τ_2	2,4	4,6	4,6
τ_3	2,4		

Table 6.4.   Pre-processed schedule table

is only required to finish after $11.6ms$, in order to meet the timing constraint. This results in an additional slack $s3$, which can be exploited to further reduce the energy dissipation. The resulting schedule, after exploiting the additional slack, is shown in Figure 6.6(c).

The pseudo-code of the voltage scaling algorithm for CTG specifications is given in Figure 6.7. We explain this algorithm in more detail next. The initial schedule table is pre-processed in step 01, so that each column corresponds to a scaling region. Practically, this involves the splitting of certain tasks, in order to distribute them over several columns. For instance, Table 6.4 shows the first three lines of the pre-processed schedule table, corresponding to the initial schedule table 6.2. As we can observe, parts of task τ_2 are additionally distributed over columns A and $\bar{A}$. This is because the condition value computed by task τ_3 becomes available after $4ms$, i.e., a scheduling decision in terms of performance (voltage) can be taken at this point.

The steps 04 to 09 apply voltage scaling to all columns in *SchTable*, in a left-to-right order. Firstly, step 04 finds all possible tracks that are followed after the condition values heading *col* are known. Additionally, out of the found tracks, the longest track (in terms of time) is identified. This track is denoted as worst case track $track_{worst}$. Depending on this worst case track, step 05 calculates the worst case slack time $slack_{worst}$, which is given by the difference between the CTG deadline Φ and the end time of the worst case track $track_{worst}$. In step 06, the slack time $slack_{col}$ distributable to column *col* is calculated by distributing $slack_{worst}$ to the columns along $track_{worst}$ in proportion to the column's duration (the difference between the latest end time and the earliest start time of the tasks in the column). Column *col* is then scaled with a *scaling-factor*, in step 07:

$$scaling\_factor = \frac{duraction_{col} + slack_{col}}{duraction_{col}} \qquad (6.2)$$

where $duration_{col}$ is the duration of column *col*. Slack on non-critical paths is exploited in step 08. And finally, step 09 has to update the contents in the columns that are subsequent to the scaled column *col* along all possible tracks.

In order to outline this scaling process in more detail, we will illustrate the function by scaling schedule table for the CTG given in Figure 6.1(a). The initial schedule table is given in Table 6.1. For simplicity we skip here step 01, since the columns already correspond to scaling regions. Let us start now with column *true*. **Step 04:** Considering that no condition value is available yet, there exist three possible tracks: track 1, track 2, and track 3 (shown in Figures 6.1(b)–(d) and Figures 6.2(a)–(c)). Since track 3 reveals the longest execution time with $26ms$ (compared to $20ms$ and $22ms$ of the tracks 1 and 2), it is marked as the worst case track. **Step 05:** Based on the timing constraint of $30ms$, the worst case slack can be calculated as $30ms - 26ms = 4ms$. **Step 06:** The slack time that is distributed to column *true* is then calculated as $4ms \cdot (10ms/26ms) = 1.538ms$, where the $10ms$ is the column's duration and $26ms$ is the worst case track execution time. **Step 07:** According to Equation (6.2), the task in column *true*, τ_1, is scaled with a factor of $(10ms+1.538ms)/10ms = 1.1538$. **Step 08:** Column *true* has no non-critical path, hence this step is skipped. **Step 09:** The successive columns of column *true*, i.e., the columns which can be executed from column *true*, are column A for track 1, columns $\bar{A}$ and $\bar{A} \wedge B$ for track 2, as well as columns $\bar{A}$ and $\bar{A} \wedge \bar{B}$ for track 3. Therefore, the schedules of these three columns require updating due to the scaling of column *true*. After applying steps 04–09, Table 6.5 is produced. Starting from the produced table and repeating the steps 04–09 for column A, Table 6.6 is generated. Here the tasks placed in column A, i.e. τ_2 and τ_7, are both extended by $4.25ms$. The final schedule table shown in Table 6.7 is generated after applying the Steps 04–09 for the columns $\bar{A}$, $\bar{A} \wedge B$, and $\bar{A} \wedge \bar{B}$, separately.

	true	A	$\bar{A}$	$\bar{A} \wedge B$	$\bar{A} \wedge \bar{B}$
τ_1	0, 11.5				
τ_2		11.5, 16.5			
τ_3			11.5, 15.5		
τ_4				15.5, 17.5	
τ_5					15.5, 21.5
τ_6				17.5, 18.5	21.5, 22.5
τ_7		16.5, 21.5		18.5, 23.5	22.5, 27.5

Table 6.5. Result after processing column *true* (values are rounded)

	true	A	$\bar{A}$	$\bar{A} \wedge B$	$\bar{A} \wedge \bar{B}$
τ_1	0, 11.5				
τ_2		11.5, 20.75			
τ_3			11.5, 15.5		
τ_4				15.5, 17.5	
τ_5					15.5, 21.5
τ_6				17.5, 18.5	21.5, 22.5
τ_7		20.75, 30		18.5, 23.5	22.5, 27.5

Table 6.6. Results after processing column A

	true	A	$\bar{A}$	$\bar{A} \wedge B$	$\bar{A} \wedge \bar{B}$
τ_1	0, 11.5				
τ_2		11.5, 20.75			
τ_3			11.5, 16.2		
τ_4				16.2, 19.6	
τ_5					16.2, 23.1
τ_6				19.6, 21.3	23.1, 24.2
τ_7		20.75, 30		21.3, 30	24.2, 30

Table 6.7. Final schedule table (scaled)

Based on the final schedule table 6.7, Figures 6.8(a)–(c) give the actual, scaled schedules of the three possible tracks of the CTG from Figure 6.1. It can be seen that these schedules meet the imposed system deadline and, at the same time, reduce the energy consumption. Furthermore, we can observe by comparing the schedules in Figure 6.3 and Figure 6.8 that the actual schedules are different from the schedule in Figure 6.3, except for track 3, which is the worst case track. It is of utmost importance to stress the fact that the schedules

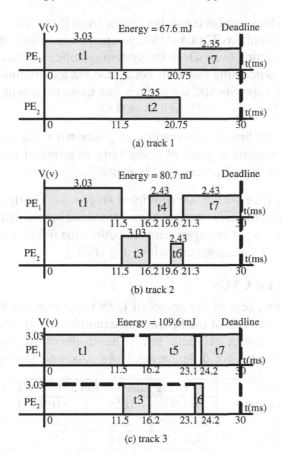

Figure 6.8.   Actual, scaled schedules

of tracks 1 and 2 in Figure 6.3 are *impracticable*! This has the following reason: The schedules shown in Figure 6.3 have been produced under the assumption that the condition values are known before actually executing the disjunction tasks. This is, of course, impossible. In reality, the condition values become available only after the disjunction tasks have finished their execution; hence, it is not possible for an on-line voltage scheduling technique to immediately use this information to achieve timing feasible and energy-efficient voltage settings.

6.4.1   Experimental Results

The DVS technique introduced in this chapter has been tested on a number of CTG examples to demonstrate its capability to produce high quality solutions in terms of energy-efficiency. The examples consist of two sets:

(a) A real-life vehicle cruise controller, taken from [50]. The CTG model of this controller contains 32 tasks, 35 edges, and 2 conditions. The underlying system architecture, onto which the system specification has been mapped, consists of 5 processing elements connected via a communication bus. In the following experiments, we assume that these processing elements are DVS-enabled ($V_{max} = 3.3V$ and $V_t = 0.8V$).

(b) The second benchmark set contains 15 generated CTG examples (ctg1–ctg15), with various degrees of complexity in terms of number of tasks, edges, conditions, and processing elements.

The following experiments are split into two groups: Firstly, we investigate the DVS technique for CTGs that has been introduced in this chapter. Secondly, we have conducted a set of experiments to outline the influence of mapping in the context of CTG system specifications and DVS.

DVS Technique for CTGs

To test the effectiveness of the proposed DVS technique, we have generated initial schedules for each of the benchmark examples and then applied the introduced voltage scaling technique to it. The results of the real-life example are shown in Table 6.8 for different deadlines. As we can observe from the

Energy dissipation before DVS (mJ)	Energy dissipation after DVS (mJ)			
	100% deadline	105% deadline	110% deadline	120% deadline
440.00	355.15	335.61	318.28	288.87

Table 6.8.   Results of the real-life example

table, without relaxing the deadline, the energy consumption is reduced from $440.00mJ$ to $355.15mJ$ when using the introduced DVS technique. This represents a reduction of 19.28%. Furthermore, by relaxing the deadline, the energy savings are increased since more slack becomes available, which can be exploited through dynamic voltage scaling. For instance, increasing the deadline by 20%, reduces the energy dissipation to $288.87mJ$, a reduction of 34.35%.

The results for the generated, hypothetical examples are given in Table 6.9. For these experiments the deadline was set to 110% of the original deadline. Similar to the previous experiment, the mapping has not been optimised in this investigation, but we consider an implicitly given mapping generated randomly together with the conditional task graph. We can observe that the introduced voltage scaling technique is capable to effectively reduce the energy dissipation of each benchmark. For instance, in the case of ctg1 the energy dissipation was reduced from $525.00mJ$ to $391.29mJ$, corresponding to a 25.5% reduc-

Benchmark example	Node/edge/condi-tion/PE number	Energy dissipation (mJ)		Reduction (%)
		before DVS	after DVS	
ctg1	13/16/2/2	525.00	391.29	25.5
ctg2	13/16/2/3	547.50	440.53	19.5
ctg3	13/16/3/2	625.00	548.12	12.3
ctg4	25/30/2/2	1475.00	1245.30	15.6
ctg5	25/30/2/4	1137.50	929.77	18.2
ctg6	25/30/3/2	1242.50	1131.11	9.0
ctg7	25/30/3/3	1413.75	1141.34	19.3
ctg8	25/29/4/2	1187.50	983.80	17.2
ctg9	35/41/2/2	1412.50	1122.18	20.6
ctg10	37/45/2/3	1803.75	1540.26	14.6
ctg11	35/41/2/5	1481.25	1191.05	19.6
ctg12	38/48/2/2	2072.50	1863.27	10.1
ctg13	42/52/2/4	2302.50	1921.13	16.6
ctg14	48/60/3/3	1845.00	1385.54	24.9
ctg15	59/71/3/3	3648.75	2998.32	17.8

Table 6.9. Results of the generated examples

tion. Similarly, the energy dissipation of ctg14 was reduced by 24.9% (from $1845.00mJ$ to $1385.54mJ$).

Please note that the calculation of the energy dissipations was based on the activation of all possible tracks given in the CTGs, i.e., an equal chance to produce a certain condition was given to all condition values. For instance, considering the CTG in Figure 6.1, track 1 has a chance to be activated of 50%, depending on condition values A and $\bar{A}$. While the tracks 2 and 3, which additionally depend on condition values B and $\bar{B}$, have both a chance of 25% to the followed. Based on these probabilities, the overall energy dissipation was calculated as a weighted average. Overall, this experiments have shown that dynamic voltage scaling can be efficiently used for CTG specification in which condition values are produced during the run-time of the application.

Mapping Optimisation

The following set of experiments demonstrates how dynamic voltage scaling for CTGs can be improved through an appropriate mapping of tasks. For this purpose the introduced voltage scaling technique was incorporated within the mapping optimisation outlined in Section 4.2. The results are given in Table 6.10, which shows the achieved energy reduction in percent and the required optimisation times in seconds (using a PentiumIII 866MHz PC). As expected,

Example	Energy reduction (%)	CPU time (s)
ctg1	38.65	10.74
ctg2	42.73	35.00
ctg3	33.56	9.82
ctg4	27.21	65.08
ctg5	31.23	143.23
ctg6	31.35	85.62
ctg7	27.69	256.40
ctg8	22.62	39.22
ctg9	30.49	14.82
ctg10	28.41	26.91
ctg11	30.68	39.15
ctg12	44.84	342.06
ctg13	50.99	1777.65
ctg14	33.22	116.34
ctg15	28.85	3639.51

Table 6.10.   Results of the mapping optimisation

the energy dissipation of each benchmark could be further reduced when the
task mapping was optimised towards the exploitation of DVS (compared to the
original scheduling technique [50]). The achieved savings go up to 50.99%
in the case of benchmark ctg13. Of course, due to the iterative optimisation
nature of the genetic algorithm, which is used for the task mapping, the optimisation times to achieve these increased savings are larger. The presented results
show how the introduced technique deals with CTGs that have up to 4 condition
values (realistic for some real-world applications). Through additional experiments have shown that CTGs with 8 conditions and 125 tasks require around 3
hours of optimisation time, which is due to the enlarged search space.

6.5   Conclusions

This chapter has demonstrated that dynamic voltage scaling can be efficiently exploited in the presence of CTG system specifications that model the
functionality of data and control dominated applications. The introduced DVS
technique exploits the slack time, taking into account the conditional behaviour
of the CTGs. Voltage and performance are scaled in such a way that deadline
constraints are fulfilled. This is ensured by considering the worst case track,
and scaling the tasks conservatively since condition values are only produced
during run-time of the application.

Notes

1 Originally called conditional process graphs.

Chapter 7

LOPOCOS: A PROTOTYPE LOW POWER CO-SYNTHESIS TOOL

Chapters 3, 4, 5, and 6 have described a number of new techniques for the design of energy-efficient distributed embedded systems. The main aim of this chapter is to show how these techniques can be used to explore the architectural design space with the intention of finding high quality system-level designs. The intention of the architectural design space exploration is the identification of a suitable architecture that will be integrated inside a new product, as outlined in Chapter 1. For this purpose the prototype co-synthesis tool, LOPOCOS (Low Power Co-Synthesis), has been developed. LOPOCOS incorporates the algorithms proposed in Chapters 3, 4, and 5. The overall goal of LOPOCOS is to equip system designers with a tool that helps to effectively explore different architectural implementations through an automated system-level design process. Using LOPOCOS, it will be demonstrated how suitable architectures (combinations of processing elements which are interconnected via communications links) can be derived for a realistic smart phone example. The smart phone example has been chosen because it is of sufficient complexity to assess the capability of LOPOCOS in terms of real-world applicability.

The remainder of this chapter is organised as follows. Section 7.1 discusses the smart phone applications and their representations as task graphs. Section 7.2 introduces LOPOCOS and demonstrates its usage for architectural design space exploration. Finally, Section 7.3 provides some concluding remarks.

7.1 Smart Phone Description

Emerging smart phones often combine cellular phones, digital cameras, and MP3 players into a single device. The specification through task graphs of such a smart phone is described in this section. The section is split into three subsections which briefly outline the applications used within the smart phone. Sec-

151

tion 7.1.1 describes the voice compression and decompression algorithms used within GSM (Global System for Mobile Telecommunication) cellular phones. Section 7.1.2 details the decompression algorithm used for the MP3 audio files. And finally, Section 7.1.3 provides information about the JPEG image compression and decompression standard. These applications and their combinations are used to model the different modes in the operational mode state machine of the smart phone (Figure 5.1, Page 101).

7.1.1 Voice Compression and Decompression

For reasonably good communications, the human voice has to be sampled at $8kHz$ and quantised with at least $13bit$. Thus, the transmission of one second of human voice requires a data transfer of $13000bytes$. To reduce this amount of data, GSM cellular phones compress voice streams using regular pulse excitation long-term predictor full-rate speech transcoders, or short RPE-LTP transcoders. These transcoders model the human voice system through two filters, the linear-predictive short-term filter and the long-term predictive filter. Figure 7.1 shows block diagrams of the encoder and the decoder units. The

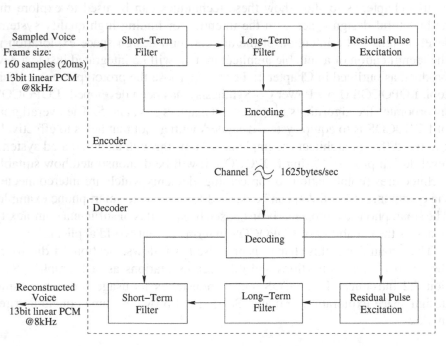

Figure 7.1. Block diagram of the GSM RPE-LTP transcoder [73]

encoder divides the incoming voice signal into short-term predictable parts, long-term predictable parts, and the remaining residual pulse. The resulting parameters of the filters and residual pulse are quantised and encoded. Using this compression technique, the $13000byte/s$ data stream is reduced to $1625byte/s$, a compression ratio of 8 to 1. Upon receiving the encoded voice stream, the decoder decompressed the parameter settings for the filters and the residual pulse. Using these settings the "original" voice signal is reconstructed. A publicly available C implementation of the GSM transcoder was developed by Degner *et al.* [4]. From this description the task graphs for the encoder and the decoder were derived, in order to permit their usage for the co-synthesis. The derivation of the task graph requires careful consideration of several aspects, including:

Loop unrolling: In order to reveal parallelism in the application, complex loops with a fixed iteration counts can be unrolled and split into several tasks.

Function collapsing: Several trivially small sub-functions are collapsed into single tasks.

Data flow extraction: Memory variables that are shared among functions are transformed into data dependencies between tasks.

Figures 7.2 and 7.3 show the extracted task graph specification of the GSM encoder and decoder. The encoder consists of 53 tasks and 80 edges, while the decoder is given by 34 tasks and 55 data dependencies. In accordance to the input frame size of 160 samples (each covering one sample period of $8kHz$), both algorithms need to be performed with a repetition rate of $20ms$. If these rate cannot be satisfied, frames are dropped, and each dropped frame corresponds to a voice gap of $20ms$, which seriously affects the quality of the communication.

In order to estimate the execution properties of each task, the specifications have been software profiled (assessment of function calls and time spend in each function) using several realistic input voice streams. The exact details about the task graph specifications and the used technology files are omitted here, due to space limits.

7.1.2   MP3 Decoder

MP3 (MPEG-1 Audio Layer 3) has become the *de facto* standard for the compression of music audio signals. Using a bit stream of $128kbps$, music can be stored in near-CD quality. The block diagram shown in Figure 7.4 outlines the functionality of the decoder. Each input frame contains a header and the signal samples. The header describes the encoding parameters such as sample frequency, stereo (mono) signal, and the block type, while the signal samples represent the encoded audio information. In the first step of the MP3 decoder, the input frames are separated into header and samples. The samples are passed to the Huffman decoder and are deqantised to derive the coefficients

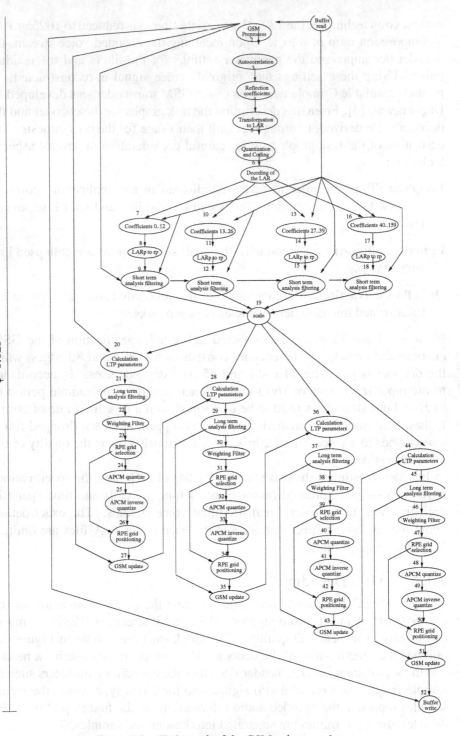

$\phi = 0.02s$

Figure 7.2. Task graph of the GSM voice encoder

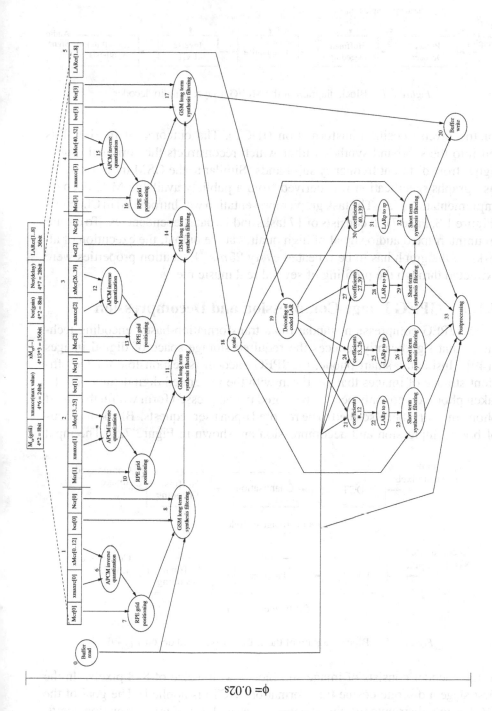

Figure 7.3. Task graph of the GSM voice decoder

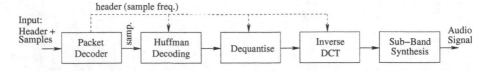

Figure 7.4. Block diagram of the MPEG-1 layer 3 audio decoder

for the inverse cosine transformation (IDCT). The outcome of the IDCT is fed into the sub-band synthesis filter which reconstructs the "original" audio signal from different frequency sub-bands. Similar to the GSM transcoder, the task graphs specification was derived from a publicly available MP3 decoder implementation [72]. The task graph representation was introduced in Chapter 1 (Figure 1.3, Page 7). It consists of 17 tasks and 18 data dependencies. To ensure an uninterrupted audio signal of high quality at the output, the execution of all tasks in the graph has to be repeated every $25ms$. Execution properties were extracted through the profiling of several real music files.

7.1.3   JPEG Image Compression and Decompression

The JPEG compression standard is a transformation-based encoding technique that significantly reduces the required storage space of digital images [149]. Inside the smart phone, the JPEG encoder is responsible for the efficient storage of images that are taken with the integrated digital camera. The taken photos do not only need to be stored in compressed form within the smart phone memory, but also need to be restored upon user requests. Block diagrams of both compression and decompression are shown in Figure 7.5. The input

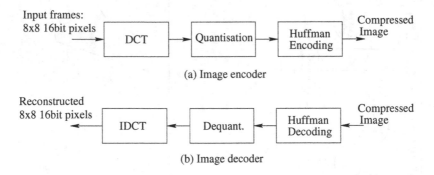

Figure 7.5. Block diagram of the JPEG encoder and decoder [149]

to the encoder consists of image sub-blocks with a size of 8x8 pixels. In the first stage a discrete cosine transformation (DCT) is applied. The goal of the DCT is to redistribute the signal energy to a small set of transformation coefficients. This allows to nullify many of the coefficients during the quantisation, which, in turn, enables an efficient encoding to reduce the storage requirements. The decoder inverts these three steps to reconstruct the image. Due to the quantisation/dequantisation this compression scheme belongs to the class of lossy encoding techniques, i.e., during the encoding some information is "lost" which results in degradation of the image quality. The JPEG library used in smart phone is also publicly available [5]. The corresponding task graphs are given in Figure 7.6, where the sequential nature of the graph can be observed. The encoder needs to store four photos per second to allow a quick series of

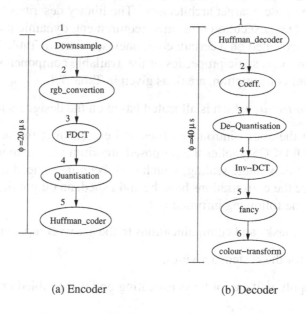

(a) Encoder   (b) Decoder

Figure 7.6.   Task graphs of the JPEG encoder and decoder

shots. Hence, the repetition rate should match $20\mu s$ for the an image size of 1024x768 pixels. The repetition rate of the decoder, on the other hand, needs to be at least $40\mu s$ in order to restore images within half a second. Similar to the GSM codec and the MP3 decoder, the JPEG encoder and decoder have been profiled on realistic input streams (i.e., photos) to extract task characteristics in terms of execution time.

7.2 LOPOCOS

LOPOCOS is a prototype co-synthesis tool that incorporates the algorithms proposed in Chapters 3, 4, and 5. The aim of this section is to show how LOPOCOS can be used for architectural design space exploration. Using the smart phone applications introduced in Section 7.1, the goal of the design space exploration is the identification of an architecture (processing elements connected through communication links) that fulfils the performance requirements and area constraints on one hand, and minimises the energy dissipation and system cost on the other. Figure 7.7 gives an overview of the design flow used within LOPOCOS. The input to LOPOCOS consists of three ASCII files, which contain the following information:

(a) *System specification* given as task graphs and operational mode state machine (in the case of multi-mode embedded systems).

(b) *Technology library* containing the available system components that can be used to compose a target architecture. The library describes the execution properties (e.g., execution time, area requirement, dynamic power) of each task when executed on a certain components, as shown Table 7.2. Furthermore, it provides static properties of the available components (e.g., price, static power consumption, area), as given in Table 7.1.

(c) *Initial architecture* which is allocated based on the designers knowledge.

After parsing the above introduced files and establishing the necessary data structures, LOPOCOS applies the proposed algorithms for mapping, scheduling, and dynamic voltage scaling, which have been introduced in Chapters 3, 4, and 5. Once the optimisations have been finished, an output file is produced that contains the following information:

(a) Mapping of tasks and communications to the active components.

(b) Schedule for the system activities.

(c) Scaled supply voltage for tasks executing on DVS-enabled processing elements.

(d) Dynamic and static power consumptions.

(e) Design quality metrics in terms of cost, area violations, and timing penalties.

(f) General information about the synthesis run, such as number of timing infeasible and area infeasible solutions produced during the optimisation.

In the case of multi-mode systems, this output is provided for each operational mode. Furthermore, LOPOCOS returns the above information as a set of Pareto optimal solutions, i.e., several not dominated solutions with different area/energy trade-offs from which the designer can choice the most suitable design.

In order to demonstrate the tool-assisted architectural design space exploration, the remainder of this section is split into two subsections. Section 7.2.1 describes briefly the necessary input files and goes into the details of the library organisation. Section 7.2.2 outlined the usage of LOPOCOS for architectural exploration.

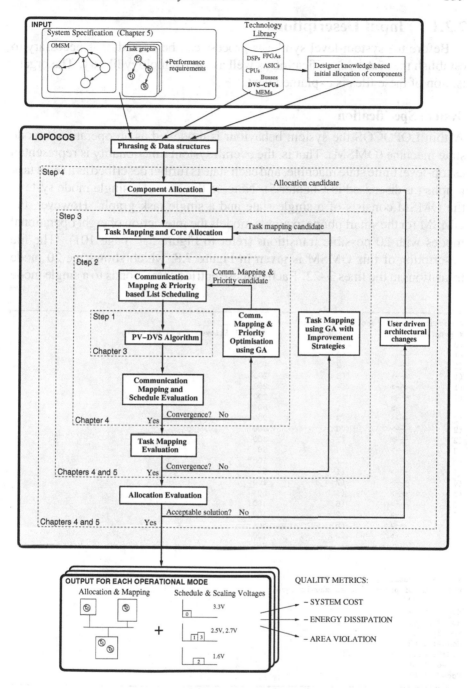

Figure 7.7. Design flow used within LOPOCOS

7.2.1   Input Descriptions

Before the system-level synthesis process can be started, it is necessary to establish the system specification as well as the technology library. The organisation of these files is explained next.

System Specification

Within LOPOCOS the system behaviour is modelled as an operational mode state machine (OMSM). That is, the overall system functionality is represented as top-level finite state machine, and each state is further described as a set of task graphs (as discussed in Chapter 5). Note, in the case of a single mode system, the OMSM consists of a single state and a single task graph. However, the OMSM for the smart phone example models the interaction of eight operational modes with 20 possible transitions (refer to Figure 5.1, page 101). The file description of this OMSM is given in Figure 7.8, which shows the 20 mode transitions in the lines 3–22. Each of these entries corresponds to a single mode

```
1    # Possible mode transitions
2    # FromMode              ToMode         Maximal Allowed Transition Time
3    (0        --->          (1             100
4    (1        --->          (0             10
5    (1        --->          (2             100
6    (1        --->          (3             200
7    (1        --->          (4             200
8    (1        --->          (5             200
9    (2        --->          (1             50
10   (2        --->          (3             200
11   (2        --->          (6             200
12   (2        --->          (7             200
13   (3        --->          (4             50
14   (3        --->          (7             100
15   (4        --->          (1             50
16   (4        --->          (7             100
17   (5        --->          (1             50
18   (5        --->          (6             100
19   (6        --->          (2             100
20   (6        --->          (5             50
21   (7        --->          (2             100
22   (7        --->          (4             50
23
24
25   # Mode execution probabilities
26   # The sum of probabilities should equal
27   Mode0: 0.09
28   Mode1: 0.74
29   Mode2: 0.01
30   Mode3: 0.02
31   Mode4: 0.02
32   Mode5: 0.10
33   Mode6: 0.01
34   Mode7: 0.01
```

Figure 7.8.   File description of the top-level finite state of the smart phone

transition and additionally states the maximal allowed time for the transition. For instance, line 7 refers to the mode transition from **Radio Link Control** (mode 1) to **Decode Photo + RLC** (mode 4), which is invoked upon a user request to restore photos, as shown in Figure 5.1. The maximal allowed time for this transition is $200ms$. Furthermore, the file shown in Figure 7.8 holds the execution probability of the eight operational modes (line 27–34), towards which the design is optimised.

In order to complete the system specification, the functionality of each operational mode has to be expressed as a task graph. An example file description of a task graph is show in Figure 7.9. This task graph represents the functionality of mode **MP3 play + RLC**. Since this mode fulfils two functions at the same time, MP3 decoding and radio link control, it contains two task graphs (lines 6–42 and line 47). The first graph represents the MP3 decoding consisting of 17 tasks and 18 edges between these tasks. This description corresponds to the specification given in Figure 1.3 on Page 7. Consider, for instance, lines 9 and 10 in Figure 7.9 which represent the **Huffman decoder** tasks τ_3 and τ_4 in Figure 1.3. The task type 40 (`ttype`) refers to a Huffman decoder, the earliest possible start time (`epst`) of the tasks is not specified, and the tasks have no imposed deadline (`dtype:NON` and `Deadline:0`). The edges between tasks are given in the lines 25–42. For example, line 29 describes the data dependency between tasks τ_3 and τ_5, i.e., the data which needs to be transfered from the Huffman decoder to the dequantisation unit. The amount of data is given by the edge type `etype`, e.g., `etype:7` refers to 256 bytes.

Technology Library

In addition to the system specification given as OMSM and task graphs, LOPO-COS requires the input of a technology library. This library contains modelling information about the available system components (processing elements and communication links). The overall goal of the co-synthesis is to appropriately select components from this library to implement the system's functionality according to the system specification. The selected components form the target architecture onto which the application is mapped and scheduled. Components modelled in LOPOCOS are general-purpose processors (GPPs), application-specific instruction-set processors (ASIPs), application-specific integrated circuits (ASICs), field-programmable gate arrays (FPGAs), communication links, and memory modules. A small size example of a technology library is shown Figure 7.10. It contains three different processing elements: an ARM7 DVS processor (lines 1–17), an ASIC in $0.35\mu m$ technology (lines 18–34), and a Xilinx Virtex II-pro FPGA (lines 35–51). Furthermore, a 16 bit wide communication bus (lines 52–58) and a standard 256kB memory module (lines 59–65) are given. These components are modelled through two data sets: *task independent component parameters* and *task dependent parameters*. The task in-

```
1   HYPERPERIOD 0.025
2   TOLERABLE_TIMING_PENALTY 1.1
3
4   # ==================MP3-DECODER=========================
5
6   Task: ( 0 0 )      ttype: 46    epst: 0    dtype: NON    Deadline: 0
7   Task: ( 0 1 )      ttype: 47    epst: 0    dtype: NON    Deadline: 0
8   Task: ( 0 2 )      ttype: 47    epst: 0    dtype: NON    Deadline: 0
9   Task: ( 0 3 )      ttype: 40    epst: 0    dtype: NON    Deadline: 0
10  Task: ( 0 4 )      ttype: 40    epst: 0    dtype: NON    Deadline: 0
11  Task: ( 0 5 )      ttype: 41    epst: 0    dtype: NON    Deadline: 0
12  Task: ( 0 6 )      ttype: 41    epst: 0    dtype: NON    Deadline: 0
13  Task: ( 0 7 )      ttype: 48    epst: 0    dtype: NON    Deadline: 0
14  Task: ( 0 8 )      ttype: 49    epst: 0    dtype: NON    Deadline: 0
15  Task: ( 0 9 )      ttype: 49    epst: 0    dtype: NON    Deadline: 0
16  Task: ( 0 10 )     ttype: 50    epst: 0    dtype: NON    Deadline: 0
17  Task: ( 0 11 )     ttype: 50    epst: 0    dtype: NON    Deadline: 0
18  Task: ( 0 12 )     ttype: 43    epst: 0    dtype: NON    Deadline: 0
19  Task: ( 0 13 )     ttype: 43    epst: 0    dtype: NON    Deadline: 0
20  Task: ( 0 14 )     ttype: 51    epst: 0    dtype: NON    Deadline: 0
21  Task: ( 0 15 )     ttype: 51    epst: 0    dtype: NON    Deadline: 0
22  Task: ( 0 16 )     ttype: 52    epst: 0    dtype: HARD   Deadline: 0.025
23
24
25  Edge: ( 0 0 )   -->  ( 0 1 )      etype: 6
26  Edge: ( 0 0 )   -->  ( 0 2 )      etype: 6
27  Edge: ( 0 1 )   -->  ( 0 3 )      etype: 6
28  Edge: ( 0 2 )   -->  ( 0 4 )      etype: 6
29  Edge: ( 0 3 )   -->  ( 0 5 )      etype: 7
30  Edge: ( 0 4 )   -->  ( 0 6 )      etype: 7
31  Edge: ( 0 5 )   -->  ( 0 7 )      etype: 7
32  Edge: ( 0 6 )   -->  ( 0 7 )      etype: 7
33  Edge: ( 0 7 )   -->  ( 0 8 )      etype: 7
34  Edge: ( 0 7 )   -->  ( 0 9 )      etype: 7
35  Edge: ( 0 8 )   -->  ( 0 10 )     etype: 7
36  Edge: ( 0 9 )   -->  ( 0 11 )     etype: 7
37  Edge: ( 0 10 )  -->  ( 0 12 )     etype: 7
38  Edge: ( 0 11 )  -->  ( 0 13 )     etype: 7
39  Edge: ( 0 12 )  -->  ( 0 14 )     etype: 7
40  Edge: ( 0 13 )  -->  ( 0 15 )     etype: 7
41  Edge: ( 0 14 )  -->  ( 0 16 )     etype: 7
42  Edge: ( 0 15 )  -->  ( 0 16 )     etype: 7
43
44
45  # ==================GSM-RLC=========================
46
47  Task: ( 1 0 )      ttype: 33    epst: 0    dtype: HARD   Deadline: 0.025
```

Figure 7.9.   File description of a single mode task graph

dependent parameters describe static values of components that are unaffected by the executed functions. With reference to Figure 7.10, the task independent parameters of the ARM7 processor are shown lines 5–8. According to the first three entries, the processor has a price of $30, a static power dissipation of $800\mu W$, and a nominal operational frequency of $80MHz$. A description of all task independent component parameters is given in Table 7.1. The task

```
1    # =====================================================
2    # ==              ** ARM7 DVS **                     ==
3    # =====================================================
4    @GPP 0 {
5    # price      StPwr(uW)    freq        pins        BEvSleep    BEvIDLE     AddsMem (32b)
6      30         800          80000       36          0.4         0.004       4294967296
7    # DVS        Vmax         vt          CommBuffer  CommTime    CommPower   CommMem
8      1          3.8          1.2         1           0.271247    25000       13
9    #-----------task properties----------------------------------------------------
10   # type  ExeCyc      DynPwr(uW)    StMem       DynMem      Preem       Exable
11     0      48820       36000         7000        3000        0           1
12     1      302300      42000         8000        2000        0           1
13     2      124040      51000         400         400         0           1
14     3      328210      38000         1600        1400        0           1
15     4      99520       37000         400         400         0           1
16     5      895280      60000         3900        1700        0           1
17   }
18   # =====================================================
19   # ==             ** AMS 0.35 **                      ==
20   # =====================================================
21   @ASIC 0 {
22   # price      StPwr(uW)    freq        pins        BEvSleep    DEvIDLE     Area(mm2)
23     60         10           2500        41          0.01        512         35.50
24   # DVS        Vmax         vt          CommBuffer  CommTime    CommPower   CommArea
25     0          2.12196      0.402969    1           0.124688    21          32
26   #-----------task properties----------------------------------------------------
27   # type ExeCyc      DynPwr(uW)    Area        Exable
28     0     222         220           12.33       1
29     1     142         141           13.20       1
30     2     40          70            4.38        1
31     3     125         324           5.49        1
32     4     146         460           6.23        1
33     5     42          60            0.93        1
34   }
35   # =====================================================
36   # ==           ** Xilinx XC2VP7 **                   ==
37   # =====================================================
38   @FPGA 0 {
39   # price      StPwr(uW)    freq        pins        BEvSleep    BEvIDLE     CLBs
40     80         127          2500        53          0.058       1759        1220
41   # DVS        Vmax         vt          CommBuffer  CommTime    CommPower   CommCLB     ReconT
42     0          2.5          0.74        1           0.4339      42.79       213         1000
43   #-----------task properties----------------------------------------------------
44   # type ExeCyc      DynPwr(uW)    CLB         Exable
45     0     288         1820          252         0
46     1     160         2590          274         0
47     2     120         910           66          0
48     3     163         2620          75          1
49     4     89          4480          128         1
50     5     298         390           11          1
51   }
52   # =====================================================
53   # ==              ** Bus 16bit **                    ==
54   # =====================================================
55   @LINK 0 {
56   # price   StPwr   freq    pins    BEvSleep    maxUser     PckSize     PckOver     aPower
57     10      775     33000   16      0.3         5           8           33          1881.648
58   }
59   # =====================================================
60   # ==          ** Memory Module 256kb **              ==
61   # =====================================================
62   @MEM 0 {
63   # price      StPwr       size        IOPwr
64     5          89.7391     256000      145.237
65   }
```

Figure 7.10. Technology library file

Abbrivation	Type	Explaination
price	GPP/FPGA/ASIC/CL/MEM	Hardware cost
StPwr	GPP/FPGA/ASIC/CL/MEM	Static power dissipation in μW
freq	GPP/FPGA/ASIC/CL	Maximal/Nominal operational frequency
pins	GPP/FPGA/ASIC/CL	Available or needed pin count for communication connection
BEvSleep	GPP/FPGA/ASIC	Break-even time from sleep mode (if implemented)
BEvIDLE	GPP/FPGA/ASIC	Break-even time from idle mode (if implemented)
AddsMem	GPP	Addressable memory size
Area	ASIC	Available area
CLBs	FPGA	Available configurable logic blocks
DVS	GPP/FPGA/ASIC	DVS enable flag (1-enabled/0-disabled)
Vmax	GPP/FPGA/ASIC	Nominal supply voltage (used for DVS)
Vt	GPP/FPGA/ASIC	Threshold voltage (used for DVS)
CommBuffer	GPP/FPGA/ASIC	Communication buffer available
CommTime	GPP/FPGA/ASIC	Bridge communication time
CommPower	GPP/FPGA/ASIC	Power dissipation during bridge communication
CommMem	GPP	Memory needed for bridge communication
CommArea	ASIC	Area needed for bridge communication
CommCLB	FPGA	CLBs needed for bridge communication
ReconT	FPGA	Clock cycles needed for a reconfiguration
ReconPwr	FPGA	Power dissipation during reconfiguration
maxUser	CL	Maximal users connected to the CL
PckSize	CL	Size of each package send over the CL
PckOver	CL	Overhead for each communication activity
aPower	CL	Average dynamic power dissipation
size	MEM	Memory size in bytes
IOPwr	MEM	Average power for memory accesses

Table 7.1. Task independent components parameters

independent parameters can be directly derived from data sheets provided by the component or core manufacturers.

The second group of parameters (task dependent parameters) describe execution properties of tasks when implemented on a certain processing element. Consider, for instance, the task independent parameters given in line 12 of the technology library (Figure 7.10), where task type 1 refers to an FFT task. Exe-

Abbreviation	Type	Explaination
type	GPP/ASIC/FPGA	refers to the task type (e.g., Huffman decoder)
ExeCyc	GPP/ASIC/FPGA	Number of clock clycles needed for execution
DynPwr	GPP/ASIC/FPGA	Dynamic power dissipation
StMem	GPP	Statically required memory
DynMem	GPP	Dynamically allocated memory
Preem	GPP	Preemption cycles (needed to store/restore registers)
Area	ASIC	Area needed for implementation
CLB	FPGA	CLBs needed for implementation
Exable	GPP/ASIC/FPGA	Feasible task/PE combination (can the task type be executed

Table 7.2. Task dependent parameters

cuting this algorithm on the ARM7 processor requires 302300 clock cycles and dissipates an average power of $42mW$.

Further, to store the task in the local memory 8000 bytes are required. In addition, 2000 bytes are allocated dynamically during run-time. The task's execution cannot be preempted. Refer to Table 7.2 for a description of the different task dependent parameters. Task dependent parameters are estimated through software and hardware profiling on realistic input data or through sophisticated estimation techniques [32, 58, 95, 144–146].

7.2.2 Architectural Design Space Exploration

A primary goal of the system designer is to find a suitable architecture that will be embedded together with the application software inside a new product. However, given the myriad of commercially available components and their possible combinations to distributed architectures, it is not hard to understand that a straight-forward architectural choice is not easy, if not impossible. The intention of LOPOCOS is to ease this design problem by equipping the designer with a tool that helps to effectively explore different architectural implementation through an automated design process. This section exemplifies the usage of LOPOCOS for tool-assisted design space exploration using the smart phone benchmark. Several different architectural choices are examined for their suitability in terms of cost, performance, area usage, and energy consumption. It is shown how the careful interpretation of the synthesis outcomes can guide the designer's architecture choices to identify solutions of high quality.

Let us consider the following design scenario. Proper product pricing as well as low energy consumption are key to a successful sale. In order to achieve the desired market success, the embedded computing system within the smart phone needs to comply with the following design constraints:

Type	em Makc	Price	Properties
GPP0	ARM7DVS-25	$20	high performance (32 bit, $25MHz$, DVS)
GPP1	ARM7DVS-10	$14	medium performance (32 bit, $10MHz$, DVS)
GPP2	6052	$5	low performance (8 bit, $6.6MHz$, DVS)
GPP3	ST68000CP10	$8	medium performance (16 bit, $10MHz$, no DVS)
ASIC0	AMS $0.6\mu m$	$30	small area ($55.2mm^2$, no DVS)
ASIC1	AMS $0.6\mu m$	$62	large area ($150mm^2$, no DVS)
ASIC2	AMS $0.6\mu m$	$42	medium area ($80mm^2$, no DVS)
ASIC3	AMS $0.6\mu m$	$68	large area ($150mm^2$, DVS)
LINK0	PCI Bus	$10	7 users, $33MHz$, 32-bit wide
LINK1	I^2C Bus	$5	1024 user, serial, $400kbps$
LINK2	CAN Bus	$7	serial, $1Mbps$
MEM0	1MB EDO RAM	$1	$100MHz$
MEM1	4MB EDO RAM	$5	$100MHz$

Table 7.3.   Components in a typical technology library

- the system price needs to be below $120 and

- the average power dissipation should not exceed $1.6mW$ (for an average user).

Of course, in addition to these constraints the architecture needs to provide a sufficiently high performance to meet the timing constraints. Certainly, at a first glance $1.6mW$ might seem unrealistically low. For a standard Lithium-Ion battery with $3.7V$ and $600mAh$, this power consumption would result in a run-time of approximately 58 days ($3.7V \cdot 0.6Ah/1.6mW \approx 1388h$). However, considering that additional analogue circuity (receiver, transmitter, audio amplifiers) as well as the display with background light consume a non-negligible amount of power, the overall battery life is reduced to several days [103]. Nevertheless, taking this into account, the digital part of the system (i.e. the embedded system) is liable for approximately 20–30% of the average power consumption, making it unavoidable to limit this power dissipation to $1.6mW$ or lower.

In order to find an architecture that fulfils the cost, power, and performance criteria, the designer can choice different components from a technology library that contains the processing elements and communication links shown in Table 7.3. We have restricted the number of components in this table; however, in reality such a technology library might contain several additional components. On top of the component names, the table gives some typical price values and component properties. Equipped with the system specification of the smart phone and the above given technology library, the designer sets out to examine different architectural choice for their suitability using LOPOCOS. The architectural design space exploration can be classified into two optimisation stages:

Performance optimisation The first goal during the design space exploration is the identification of such architectures that fulfil performance requirements of the application without penalties in area and cost.

Energy Minimisation Once viable architectural implementations have been established, the next stage in the optimisation hierarchy is the energy minimisation. Based on the design knowledge gathered during the performance optimisation, the designer aims to tune the architecture towards energy efficiency by exploiting DVS and by carefully increasing the available hardware area to allow the implementation of energy critical tasks in hardware.

The following considers both optimisation steps.

7.2.2.1 Performance Optimisation

<u>Architecture 1:</u> The purpose of the performance optimisation is the identification of architectures that fulfil the computational requirements of the smart phone specification. In accordance to the design cost and performance constraints, the designer initially allocates components based on intuition or previous-design experience. In fact, many new system designs are upgrades from older products. Upgrading existing systems can significantly cut down the design costs. Nevertheless, the increasing complexity of new applications demands the re-evaluation of the architecture's suitability. For instance, a previously designed GSM cellular phone is implemented on an architecture that consists of a low performance 8-bit CPU (including a 1MB RAM) and an ASIC with an usable area of $55.2mm^2$, both interconnected via an I^2C bus. The resulting price of this configuration is $41, according to the components cost given in Table 7.3. Considering the GSM transcoder only, the architecture provided sufficiently high performance and even resulted in some performance headroom. To evaluate if this performance headroom is large enough to additionally perform the MP3 decoder as well as the JPEG compression/decompression the designer applies LOPOCOS. Figure 7.11 shows the co-synthesis results for this architecture as a trade-off between average power dissipation and area penalty. Each diamond represents a timing feasible implementation, that is, an implementation that satisfies the imposed deadlines and repetition rates of the individual smart phone applications. Consider, for instance, the rightmost implementation candidate which has an average power dissipation of $2.5mW$. This solution shows an area penalty of 2.6, i.e., the available hardware ($55.2mm^2$) is exceeded by 2.6 times. As the figure illustrates, all solution candidates found violate the area constraints (area penalty > 1), hence, the allocated architecture does not provide sufficient hardware area to accommodate the timing critical

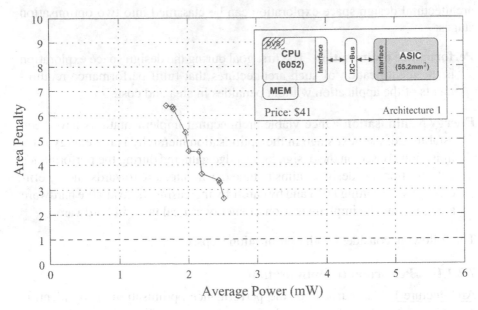

Figure 7.11. Co-synthesis results of Architecture 1

tasks of the smart phone example. Hence, to viable implement this solution, a hardware area of at least $143.5mm^2$ ($55.2mm^2 \cdot 2.6$) would be necessary. On the other hand, using a fast processor instead of the low performance 6052 CPU would allow to move some timing critical tasks in the design towards software implementations, and therefore reduce the required area.

Clearly, increasing the hardware area (larger ASIC) as well as allocating a fast processor would increase the system cost. In the following both possibilities are examined.

Architecture 2: In the first possibility to improve the system design, the designer aims to fulfil the hardware requirements of at least $143.5mm^2$. This can be achieved by replacing the current ASIC 0 with an area of $55.2mm^2$ by an ASIC 3 which offers $150mm^2$ (Table 7.3). In this way, the area requirements of the smart phone specification should be satisfied. However, the large ASIC increases the price of the embedded system to $79. The co-synthesis outcome of LOPOCOS for this configuration is illustrated in Figure 7.12 (indicated as Architecture 2). As expected, this architecture offers sufficient area and computational power to perform all applications in the smart phone device adequately. This is indicated by the rightmost solution with an area penalty of 1 (no area violation) and an average power dissipation of $2.1mW$.

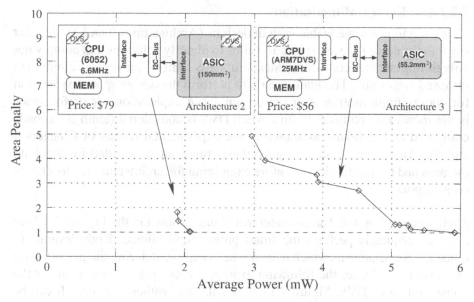

Figure 7.12. Co-synthesis results of Architectures 2 and 3

Architecture 3: Let us consider the second possible modification of the initial allocation. Instead of exchanging ASIC0 with ASIC3, the 6052 8-bit CPU (6.6MHz) is replaced with a more powerful ARM7DVS (25MHz). The price for this implementation is $56. The architecture together with the co-synthesis results are shown in Figure 7.12 (indicated as Architecture 3). As it can be observed from the figure, this combination of components allows to viably implement the applications. This is achieved due to the higher performance of the ARM7 compared to the 6052 CPU, which allows to run more of the timing critical tasks in software. Yet, the average power dissipation for the solution with no area penalty (represented rightmost diamond) is close to 6mW, which exceeds the permitted value of 1.6mW by nearly 4 times. In fact, also Architecture 2, which achieved an average power dissipation of 2.1mW by increasing the hardware area, could not satisfy the imposed maximal power dissipation of 1.6mW.

In summary, the design space exploration carried out so far, concentrated on area and performance aspects only. Starting from an initial yet not viable architecture (Architecture 1), the design was refined in two possible directions. Firstly, by increasing the available hardware (Architecture 2), and secondly, by increase the performance of the software processor. In both case viable designs could be synthesised with LOPOCOS, however, none of which could satisfy the imposed power constraint. The aim of the following design space exploration is to further refine the system design in order to meet the power constraint.

7.2.2.2   Energy Minimisation

Having found some viable system designs (Architectures 2 and 3), the next stage in the design space exploration aims to identify solutions that comply not only timing and area constrains, but additionally limit the power consumption to imposed constraint. The main reason to perform the design space exploration for energy minimisation after the design space exploration for performance is the increased optimisation time when DVS is considered (tenth of seconds compared to 1–2 hours, as shown in the experimental results of Chapters 4 and 5). Thus, unnecessary long synthesis runs can be avoided by checking the area and timing feasibility before examining the architecture for its energy consumption.

Architectures 2 and 3: The architectures 2 and 3 shown in the Figures 7.12 are able to adequately perform the smart phone applications. Both designs include DVS components which have not been exploited during the performance optimisation. Hence, the following examines both system designs under the consideration of DVS. Figure 7.14 shows the co-synthesis results. It can be

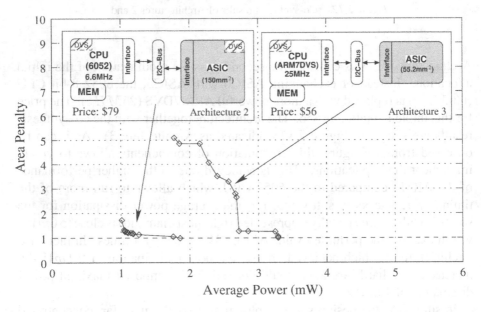

Figure 7.13.   Co-synthesis results for Architectures 2 and 3, exploiting DVS

observed that in both cases the average power dissipation was reduced, as a comparison between Figures 7.12 and 7.13 reveals. For instance, the power dissipation of Architecture 2 was reduced from $2.1mW$ to $1.9mW$. Similarly, the power dissipation of Architecture 3 was reduced from $5.9mW$ to $3.3mW$. Nevertheless, both architectures still exceed the imposed maximal power dissipation of $1.6mW$. To overcome this problem, a further design refinement is required.

<u>Architectures 4:</u> In general, hardware implementations of tasks are 1–2 orders of magnitude more energy efficient than software implementations [35]. Hence, one possible way to reduce the power dissipation is to map more tasks to hardware. Clearly, this necessitates to increase the available hardware area. In a similar manner, increasing the performance of software-programmable can help to reduce energy. This is achieved due to the fact that a fast processor is potentially able to satisfy the performance needs of timing critical tasks that otherwise would require a hardware implementation. Thus, more hardware area can be used to efficiently implement task with high energy dissipation. Of course, faster processor consume more power than their slower counterparts, hence the design needs to be carefully balanced. Consider, for example, the target architecture shown in Figure 7.14. This design consists of an ARM7DVS

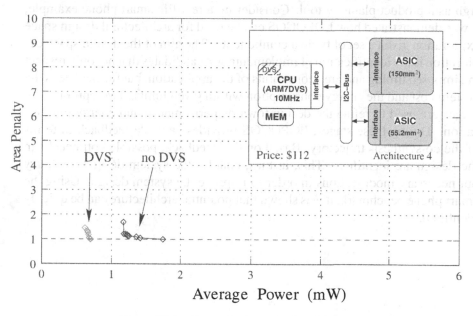

Figure 7.14. Co-synthesis results for Architecture 4

processor operating at $10MHz$ and two ASIC which provide areas of $150mm^2$ and $55.2mm^2$. All components are connected via a single I²C bus. The total cost of this architecture is \$112, which still complies the price limit of \$120. Figure 7.14 shows the co-synthesis results for the architecture, with and without the consideration of DVS. As it can be seen from the figure, if the DVS feature of ARM7 processor is not exploited, an average power dissipation of $1.7mW$ can be achieved. This is still higher than the imposed power constraint of $1.6mW$. However, optimising the implementation under consideration of DVS, the average power is reduced to $0.7mW$. This is the first implementation that meets all imposed design constraints. The cost is below \$120, the timing constraints and area limitations are not violated, and the dissipated average power is $0.7mW$ when using DVS.

It should be noted that this is only one possible solution. Nevertheless, the system designer might aim to find even better solutions by further exploring the architectural design space. Finding the presented solution took approximately 4.5 hours on a PentiumIII/$1.2GHz$ PC.

7.3   Concluding Remarks

The architectural design space of distributed embedded systems is vast. Effective design tools are essential to support the system designers in exploring this solution space quickly and thoroughly. This chapter has introduced LOPO-COS, a prototype co-synthesis tool that can be used for high-level system design as a product planning tool. Considering a real-life smart phone example, it was demonstrated how LOPOCOS can be used for architectural design space exploration given a set of typical components. The aim of the design space exploration is the identification of implementation candidates that not only respect the imposed timing and area constraints of the application, but further keeps the system cost and the energy consumption within given limits. Two optimisation stages are used to refine the design towards performance and energy minimisation. During these stages, LOPOCOS provides valuable feedback in terms of the co-synthesis trajectory. Based on this feedback, possible bottlenecks in the design are identified. This, in turn, enables the system designer to carry out necessary modifications in order to improve the system design. Using the smart phone benchmark, it was shown that potential architecture can be quickly identified.

Chapter 8

CONCLUSION

It is very likely that the demand for energy-efficient portable systems, as well as their complexity, particularly in consumer electronics, will continue to increase. Computational intensive applications such as multimedia with advanced audio and video coding techniques, which, until recently, were only practicable on high-performance stationary workstations, are entering the mobile arena. Designing energy-efficient embedded computing systems for such applications is a challenging and difficult task. This is because of the tight energy budget of the powering batteries as well as the massive pressure of the consumer electronics market with shortening design cycles and low product costs constraints which are essential for the success of these emerging and next-generation products. The work presented in this book has focused on the design of energy-efficient single-mode and multi-mode distributed embedded systems. To achieve energy-efficient designs, novel techniques and algorithms have been developed that can be used within co-synthesis frameworks to automated the design process of such embedded systems. In particular three key issues have been addressed:

(a) *Dynamic voltage scaling* was investigated in the context of single-mode distributed heterogeneous embedded computing systems. For this purpose a number of algorithms have been proposed. A new power variation-driven voltage scaling technique, based on a novel energy-gradient strategy, which overcomes restrictions associated with traditional approaches, has been introduced. In addition, voltage scaling for conditional task graph specifications was outlined. New co-synthesis techniques for scheduling and mapping have been presented, which were specifically developed for an effective utilisation of DVS by carefully increasing available slack time on DVS-enabled processing elements.

173

(b) The design of *energy-efficient multi-mode embedded computing systems* has been investigated for the first time and a novel co-synthesis methodology, which enables an effective sharing of limited hardware resources by considering mode execution probabilities, has been introduced.

(c) LOPOCOS, a prototype co-synthesis tool that incorporates the techniques and algorithms proposed in this book, has been developed, with the aim to extensively automate the system-level design process of energy-efficient embedded system. The automation of the design process enables an effective architectural design space exploration, which supports the system designer in finding a target architecture for high quality system implementations.

The following Section 8.1 will summerise the introduced techniques, and Section 8.2 will outline some future directions in the area of low power synthesis.

8.1   Summary

Dynamically voltage-scalable processors, which are capable of rapidly changing their operational state in terms of voltage and frequency setting "on-the-fly" in order to trade off energy against performance, have recently become available. The work presented in this book has focused on the energy minimisation of such distributed embedded systems that contain DVS processors.

Chapter 3 has introduced a new power variation-driven voltage scaling technique which takes into account the individual power dissipated by tasks during the voltage selection, in order to increase the achievable energy savings. A energy-gradient based heuristic has been developed for this purpose, which uses a local energy measurement, the energy-gradient, to scale tasks that potentially lead to high energy-efficiency. Using a mapped-and-scheduled task graph structure (MSTG) allows a propagation of execution time changes throughout the schedule in linear time; important for a fast execution of the scaling algorithm. Furthermore, the influence of a minimal extension time was investigated with regard to energy reduction possibility and performance. Experimental results have shown that up to 38% savings can be achieved for small real-life benchmarks compared to a more simple and restricted approach that uses a fixed power model.

In order to further reduce the energy dissipation of embedded systems, a system-level co-synthesis technique particular tailored for DVS was introduced in Chapter 4. By appropriately mapping and scheduling the system tasks and communications, the energy reduction capability of DVS is effectively exploited. This is achieved through a thorough mapping and scheduling optimisation based on a two-step genetic algorithm. In the first step, the schedule of system activities (tasks and communications) as well as the communication mapping are optimised not only towards timing feasibility but additionally towards energy minimisation through DVS. This is done using a genetic algorithm with a specialised genome representation that combines schedule and communication mapping into a single string. The second step is responsible for the mapping of tasks onto the processing elements of the target architecture, carefully balancing the design between area feasibility and energy minimisation. Both optimisation steps are guided by an energy estimation based on the voltage scaling technique introduced in Chapter 3. In this two-step approach, task mapping and communication mapping have been separated to avoid creation of infeasible communication mappings. As a result, the search space is restricted to structurally feasible mapping solutions only, reducing the optimisation time while maintaining the full possible optimisation potential. Extensive experiments have been conducted including a real-life benchmark of an optical flow detection algorithm. These experiments have shown that appropriate mapping and scheduling is essential in order to achieve high energy savings and simultaneously timing feasible solutions. The proposed two-step approach based on genetic algorithms accomplished this in moderate optimisation times.

Chapters 3 and 4 focused on the energy minimisation of single-mode systems. The co-synthesis approach introduced in Chapter 5 addressed the design of energy-efficient distributed embedded systems that need to perform multimodes. Although many embedded systems often perform single specific functionalities, such a standalone MP3 player, modern and next-generation mobile devices demand the support for multiple operational modes, i.e., the operation across a set of several different applications. The fundamental difference between single mode and multi-mode systems is that multi-mode systems have the possibility to share computational resources among the individual modes of operation. Thereby, an effective system-level co-synthesis technique, which accounts for resource sharing, can help to significantly reduce hardware cost. The multi-mode co-synthesis approach introduced in Chapter 5 is based on a new system specification model called operational mode state machine (OMSM), which captures mode interaction together with mode functionality. Considerable energy saving are achieved by taking into account the execution probabilities of the different modes during the co-synthesis process. These savings are possible even without the employment of DVS. The main problem within this synthesis process was the development of a new mapping strategy which considers the uneven execution probabilities. This mapping strategy allows the multiple task implementation on different processing elements, whenever such multiple implementations reduce the energy consumption without significantly increasing the system cost. The mapping optimisation itself is based on a genetic algorithm. Four new improvement strategies help to improve this optimisation process in terms of run-time and solution quality. This is achieved by carefully pushing the optimisation into promising search space regions. Additionally, dynamic voltage scaling was investigated in the context of multi-mode system. A virtual transformation of tasks executing on DVS-enabled hardware was shown to be useful in the presence of simultaneously scalable cores which potentially execute tasks in parallel. Overall, numerous experiments clearly indicate the advantage of taking mode execution probabilities into account during the co-synthesis.

In Chapter 6, we have concentrated on dynamic voltage scaling in the context of applications that exhibit extensive control flow. The introduced voltage scaling technique for conditional task graph specifications identifies the worst case track and produces a scaled schedule table that ensures the satisfaction of the imposed deadline, while, at the same time, reduces the energy consumption. In accordance to condition values, which are produced during run-time of the application, the suitably scaled voltages are applied based on the scaled schedule table. Additional slack on non-critical paths is exploited using the techniques of Chapter 3. Several benchmark experiments were conducted and the outcomes have illustrate the efficiency of DVS for systems represents through conditional task graphs.

Finally, Chapter 7 has introduced LOPOCOS, a prototype system-level co-synthesis tool which incorporates the techniques and algorithms proposed in Chapters 3, 4, and 5. The tool aids the system designer in finding system implementations that fulfil the imposed design constraints. Using a real-life smart phone example, a combination of cellular phone, MP3 player, and digital camera, it was outlined how LOPOCOS can help to effectively explore the architectural design space with the aim of finding area and timing feasible implementation that minimise system cost and energy consumption, simultaneously.

In conclusion, the broad aim of this book was to introduce automated design techniques for energy-efficient distributed embedded systems. A detailed investigation into dynamic voltage scaling for distributed embedded systems has shown that substantial energy savings can be achieved through carefully ordering the execution of activities as well as through an appropriate mapping of the application on the target architecture. Furthermore, it was shown that the consideration of the mode execution probabilities (activation times) during the co-synthesis of multi-mode embedded systems is essential for an appropriate resource sharing that yields energy-efficient designs. Considering a realistic smart phone example it was demonstrated that up to 66% in energy savings can be achieved compared to a naive approach that neglects execution probabilities as well as dynamic voltage scaling. In essence, when seeking to design energy-efficient distributed embedded systems, system-level co-synthesis techniques that consider energy management techniques and mode execution probabilities (as the approaches presented in this book) should be given serious consideration.

8.2   Future Directions

The work outlined in this book could be elaborated in many different directions. The following introduces some relevant areas that will become of interested in the future.

8.2.1   Leakage Power Reduction

Given the constantly shrinking feature sizes of silicon technology, single chip systems will be soon accommodate billions of transistors [127]. In order to reduce the active power consumption of these systems, the nominal supply voltage is reduced with every new technology generations. Nevertheless, the reduction of the nominal supply voltage levels requires an additional down-scaling of the threshold voltage to maintain high operational frequencies. This lower threshold voltage comes at the cost of an increased leakage power consumption, which will become comparable to the active (dynamic) power consumption in the near future. For instance, the leakage power consumption of an inverter circuit in the predictive $70nm$ technology (operating at $125°C$) accounts for approximately 50% of the total power dissipation [86]. While dynamic voltage scaling is effective in reducing the dynamic power consumption, it is less efficient in limiting the leakage power. One approach to address the leakage power is adaptive body biasing [86, 102]. Similar to dynamic voltage scaling, which changes the supply voltage during run-time of an application, it is possible to scale the transistors' body bias voltage during operation. Thereby, allowing to increasing the threshold voltage of the circuit at the cost of a lower performance. In this context, new system-level synthesis techniques are required that allow the consideration of a concurrent adaption of the circuit's supply voltage as well as its threshold voltage, in order to find a good design trade-off between total power consumption (leakage and dynamic) and performance.

8.2.2   Energy-aware Granularity Selection

The level of granularity has an influence on the performance optimisation potential of the application as well as on the optimisation time of the co-synthesis process [75]. Nevertheless, when designing energy-efficient systems, the selected granularity can have an additional effect on the energy aspects of the implementation possibilities. Consider, for instance, a single task that can be split into two separate smaller tasks. One of which is active for most of the computations of the initial task, while the other requires only short execution time. Clearly, when optimising the mapping it is likely that the initial (single) task is placed into hardware due to the lower execution time and power reduction (correspondingly, low energy consumption). However, in the case of the split task version, a refined mapping can place the task with high computational overhead into hardware, while the less computational expensive task part is placed into software, avoiding unnecessary waste of hardware resources. Accordingly, the development of an energy-aware granularity selection scheme could further improve the co-synthesis process.

8.2.3   Platform Design

As the complexity of embedded systems and underlying hardware continues to increase, the design cost for these have been risen drastically. For instance, a single mask set for a state-of-the-art chip exceed a cost of $0.5M. With shrinking feature size this cost will further increase [6, 127]. To cope this tremendous cost, the design of next-generation system-on-a-chip (SoC) embedded systems will be most likely based on platforms, i.e., pre-designed hardware architectures that can be used over a set of related applications [85]. Even if this hardware might be over-designed for the particular application at hand. Nevertheless, a challenging problem is the identification of platforms that allow a most cost-effective implementation of several different applications. Not only cost-effectiveness is essential here. Also energy consumption is an aspect that needs careful consideration when seeking to design a widely usable platform. System-level co-synthesis can help during the difficult choice of an appropriate platform. For instance, a design team might be confronted with the design of three different, yet related application (e.g., a smart phone, a PDA, and a portable MP3/DVD player). The development time and cost of these products could be significantly reduced if one would find a single hardware architecture suitable for all three applications. This platform might be over-designed to a certain extend depending on the application that needs to be implemented, but the expensive and time consuming hardware design is reduced significantly to a single design process. Here investigations into concurrent co-design methodology for several different applications are needed. Ultimately, the development of of novel "platform-level" co-synthesis techniques that consider emerging paradigms, such as networks-on-chip, can aid the designers in finding a good hardware platform.

8.2.4   Networks on Chip

In the last decade, component reuse has turned out to be one of the most promising design methodologies to overcome the ever-increasing design productivity gap. This is achieved by building new systems-on-single-chips (SoC) out of pre-designed hardware components, in a similar fashion as software functions are reused within different applications. Nevertheless, the enormous advances in silicon technology have enabled the design of systems with several dozens or hundreds of communicating components residing on the same die. In order to interconnect these components globally, network-like communication infrastructures are essential, since long, single bus architectures become

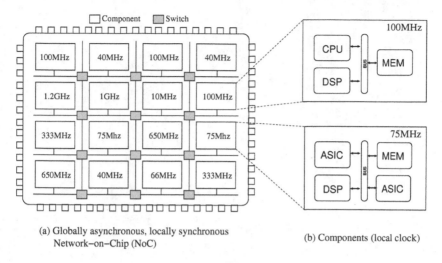

(a) Globally asynchronous, locally synchronous Network–on–Chip (NoC)

(b) Components (local clock)

Figure 8.1.

inefficient. Such networked systems are today referred to as *Network-on-Chip (NoC)* [27, 81]. A typical NoC is shown in Figure 8.1, which consists of several components operating at different clocks. Communication between these multiple clock domains is performed over a global network, while intra-component communication takes place over fast, local links, such as buses. This *globally asynchronous, locally synchronous* paradigm is necessary since the transmission of global signals will soon require several clock cycles to traverse the whole chip, making it complicated to avoid clock screw.

Certainly, to fully exploit the potential of these next-generation systems, novel system design methodologies have to be developed. In particular, problems such as power-aware routing, control protocols for reliable inter-component communications over unreliable links, and the intrinsic non-determinism of communication performance have to be addressed. Furthermore, the embedded operation systems, which will run on the NoC architectures, need further sophistication in order to enable the effective usage of energy management strategies, considering both communicating and computing power consumption when taking on-line decisions.

References

[1] GNU CC Manual.
    available at: http://gcc.gnu.org/.

[2] GNU GCJ: GNU Compiler for Java.
    available at: http://gcc.gnu.org/java/.

[3] GNU gprof Manual.
    available at: http://www.gnu.org/manual/gprof-2.9.1/gprof.html.

[4] GSM 06.10, Technical University of Berlin.
    Source code available at http://kbs.cs.tu-berlin.de/~jutta/toast.html.

[5] Independent JPEG Group: jpeg-6b.
    Source code available at ftp://ftp.uu.net/graphics/jpeg/jpegsrc.v6b.tar.gz.

[6] International Technology Roadmap for Semiconductors 2000.
    http://public.itrs.net.

[7] Standard Performance Evaluation Corporation.
    http://www.specbench.org/.

[8] Synopsys Behavioral Compiler.
    http://www.synopsys.com/products/beh_syn/beh_syn.html.

[9] Synopsys Design Compiler.
    http://www.synopsys.com/products/logic/design_comp_ds.html.

[10] Synplify from Synplicity.
    http://www.synplicity.com/products/.

[11] The Free-IP Project.
    http://www.free-ip.com/.

[12] WITAS: The Wallenberg laboratory for research on Information Technology and Autonomous System.
    http://www.ida.liu.se/ext/witas/.

[13] Intel® XScale™ Core, Developer's Manual, December 2000. Order Number 273473-001.

[14] Mobile AMD Athlon™4, Processor Model 6 CPGA Data Sheet, November 2000. Publication No 24319 Rev E.

[15] Intel® PXA800F Cellular Processor, Developer's Manual, February 2003. Order Number 252569-002.

[16] T. Adam, K. Chandy, and J. Dickson. A Comparison of List Scheduling for Parallel Processing Systems. *J. Communications of the ACM*, 17(12):685–690, December 1974.

[17] Alexandru Andrei, Marcus T. Schmitz, Bashir M. Al-Hashimi, and Petru Eles. Overhead-Conscious Voltage Selection for Dynamic and Leakage Energy Reduction of Time-Constrained Systems. In *Proceedings Design, Automation and Test in Europe Conference (DATE2004)*, February 2004.

[18] Thomas Bäck, Ulrich Hammel, and Hans-Paul Schwefel. Evolutionary Computation: Comments on the History and Current State. *IEEE Transactions on Evolutionary Computation*, 1(1):3–17, 1997.

[19] Z. Baidas, A. D. Brown, and A. C. Williams. Floting-point Behavioral Synthesis. *IEEE Transactions on Computer-Aided Design*, 20(7):828–839, July 2001.

[20] N. Bambha, S. Bhattacharyya, J. Teich, and E. Zitzler. Hybrid Global/Local Search Strategies for Dynamic Voltage Scaling in Embedded Multiprocessors. In *Proceedings 1st International Symposium Hardware/Software Co-Design (CODES'01)*, pages 243–248, April 2001.

[21] Armin Bender. Design of an Optimal Loosely Coupled Heterogeneous Multiprocessor System. In *Proceedings European Design Automation Conference*, pages 275–281, March 1996.

[22] L. Benini, G. De Micheli, E. Macii, D. Sciuto, and C. Silvano. Address Bus Encoding Techniques for System-Level Power Optimization. In *Proceedings Design, Automation and Test in Europe Conference (DATE98)*, pages 861–866, March 1998.

[23] Luca Benini, A. Bogliolo, and Giovanni De Micheli. A Survey of Design Techniques for System-Level Dynamic Power Management. *IEEE Transactions on VLSI Systems*, pages 299–316, June 2000.

[24] Luca Benini, A. Bogliolo, G.A. Paleologo, and G. De Micheli. Policy Optimization for Dynamic Power Management. *IEEE Transactions on Computer-Aided Design*, 18(6):813–833, June 1999.

[25] Luca Benini, Alessandro Bogliolo, and Giovanni De Mecheli. Dynamic Power Management of Electronic Systems. In *Proceedings IEEE/ACM International Conference Computer-Aided Design (ICCAD-98)*, pages 696–702, Nov 1998.

[26] Luca Benini and Giovanni De Micheli. *Dynamic Power Management: Design Techniques and CAD Tools*. Kluwer Academic Publishers, 1997.

[27] Luca Benini and Giovanni De Micheli. Networks on Chips: A New SoC Paradigm. *IEEE Computer*, 35(1):70–78, January 2002.

[28] Luca Benini, Giovanni De Micheli, Enrico Macii, Massimo Poncino, and R. Scarsi. Symbolic Synthesis of Clock-gating Logic for Power Optimization of Synchronous Controllers. *ACM Transactions on Design Automation of Electronic Systems*, 4(4):351–375, 1999.

[29] Luca Benini, Polly Siegel, and Giovanni De Micheli. Saving Power by Synthesizing Gated Clocks for Sequential Circuits. *IEEE Design & Test of Computers*, 11(4):32–41, 1994.

[30] Peter Bjørn-Jørgensen and Jan Madsen. Critical Path Driven Cosynthesis for Heterogeneous Target Architectures. In *Proceedings 5th International Workshop Hardware/Software Co-Design (Codes/CASHE'97)*, pages 15 – 19, 1997.

[31] Shekhar Borkar. Design Challenges of Technology Scaling. *IEEE Mirco*, pages 23–29, July 1999.

[32] C. Brandolese, W. Fornaciari, F. Salice, and D. Sciuto. Energy Estimation for 32 bit Microprocessors. In *Proceedings 8th International Workshop Hardware/Software Co-Design (CODES'00)*, pages 24–28, May 2000.

[33] Jason J. Brown, Danny Z.Chen, Garrison W. Greenwood, Xiaobo (Sharon) Hu, and Richard W. Taylor. Scheduling for Power Reduction in a Real-Time System. In *Proceedings International Symposium Low Power Electronics and Design (ISLPED'97)*, pages 84–87, 1997.

[34] Thomas D. Burd. *Energy-Efficient Processor System Design*. PhD thesis, University of California at Berkeley, 2001.

[35] Thomas D. Burd and Robert W. Brodersen. Processor Design for Portable Systems. *Journal on VLSI Signal Processing*, 13(2):203–222, August 1996.

[36] Thomas D. Burd, Trevor A. Pering, Anthony J. Stratakos, and Robert W. Brodersen. A Dynamic Voltage Scaled Microprocessor System. *IEEE Journal on Solid-State Circuits*, 35(11):1571–1580, November 2000.

[37] Anantha P. Chandrakasan and Robert W. Brodersen. *Low Power Digital CMOS Design*. Kluwer Academic Publisher, 1995.

[38] Anantha P. Chandrakasan, T. Sheng, and Robert W. Brodersen. Low Power CMOS Digital Design. *Journal of Solid State Circuits*, 27(4):473–484, April 1992.

[39] Eui-Young Chung, Luca Benini, and Giovanni De Micheli. Contents Provider-Assisted Dynamic Voltage Scaling for Low Energy Multimedia Applications. In *Proceedings International Symposium Low Power Electronics and Design (ISLPED'02)*, pages 42–47, August 2002.

[40] J. D'Ambrosio and X. Hu. Configuration-Level Hardware/Software Partitioning for Real-Time Embedded Systems. In *Proceedings International Workshop Hardware/Software Co-Design (Codes/CASHE'94)*, pages 34–41, 1994.

[41] Bharat P. Dave, Ganesh Lakshminarayana, and Niraj K. Jha. COSYN: Hardware-Software Co-Synthesis of Embedded Systems. In *Proceedings IEEE 34th Design Automation Conference (DAC97)*, pages 703–708, 1997.

[42] G. DeMicheli. *Synthesis and Optimization of Digital Circuits.* McGraw-Hill, 1994.

[43] Giovanni DeMicheli, David C. Ku, Frederic Mailhot, and Thomas Truong. The Olympus Synthesis System for Digital Design. *IEEE Design & Test of Computers,* pages 37–53, October 1990.

[44] Micheal L. Dertouzos and Aloysius Ka-Lau Mok. Multiprocessor On-Line Scheduling of Hard-Real-Time Tasks. *IEEE Transactions on Software Engineering,* 15(12):1497–1506, December 1989.

[45] Srinivas Devadas and Sharad Malik. A Survey of Optimization Techniques Targeting Low Power VLSI Circuits. In *Proceedings IEEE 32nd Design Automation Conference (DAC95),* pages 242–247, 1995.

[46] Muhammad K. Dhodhi, Imtiaz Ahmad, and Robert Storer. SHEMUS: Synthesis of Heterogeneous Multiprocessor Systems. *J. Microprocessors and Microsystems,* 19(6):311–319, August 1995.

[47] R. Dick and N. K. Jha. MOCSYN: Multiobjective core-based single-chip system synthesis. In *Proceedings Design, Automation and Test in Europe Conference (DATE99),* pages 263–270, March 1999.

[48] R. Dick, D. Rhodes, and W. Wolf. TGFF: Task Graphs for free. In *Proceedings 5th International Workshop Hardware/Software Co-Design (Codes/CASHE'97),* pages 97–101, March 1998.

[49] Robert P. Dick and Niraj K. Jha. MOGAC: A Multiobjective Genetic Algorithm for Hardware-Software Co-Synthesis of Distributed Embedded Systems. *IEEE Transactions on Computer-Aided Design,* 17(10):920–935, Oct 1998.

[50] Petru Eles, Alexa Doboli, Paul Pop, and Zebo Peng. Scheduling with Bus Access Optimization for Distributed Embedded Systems. *IEEE Transactions on VLSI Systems,* 8(5):472–491, Oct 2000.

[51] Petru Eles, Krzysztof Kuchcinski, Zebo Peng, Alexa Doboli, and Paul Pop. Scheduling of Conditional Process Graphs for the Synthesis of Embedded Systems. In *Proceedings Design, Automation and Test in Europe Conference (DATE98),* pages 132–138, 1998.

[52] Petru Eles, Zebo Peng, Krzysztof Kuchcinski, and Alexa Doboli. System Level Hardware/Software Partitioning Based on Simulated Annealing and Tabu Search. *Kluwer Journal on Design Automation for Embedded Systems,* 2:5–32, 1997.

[53] R. Ernst and J. Henkel. Hardware-Software Codesign of Embedded Controllers Based on Hardware Extraction. In *1st International Workshop Hardware/Software Co-Design (Codes/CASHE'92),* 1992.

[54] R. Ernst, J. Henkel, and Th. Brenner. Hardware-Software Co-synthesis for Mirco-Controllers. *IEEE Design & Test of Computers,* 10(4):64–75, Dec 1993.

[55] Rolf Ernst. Codesign of Embedded Systems: Status and Trends. *IEEE Design & Test of Computers,* pages 45–54, April 1998.

[56] A. Feller. Automatic layout of low-cost quick-turnaround random-logic custom LSI devices. In *Proceedings IEEE 13th Design Automation Conference (DAC76),* pages 97–85, 1976.

[57] Terence C. Fogarty. Varying the probability of mutation in the genetic algorithm. In *Proceedings 3rd International Conference on Genetic Algorithms (ICGA)*, pages 104–109, 1989.

[58] W. Fornaciari, D. Sciuto, and C. Silvano. Power Estimation for Architectural Exploration of HW/SW Communication on System-Level Buses. In *Proceedings 7th International Workshop Hardware/Software Co-Design (CODES'99)*, pages 152–156, May 1999.

[59] Michael L. Fredman and Robert Endre Tarjan. Fibonacci Heaps and Their Uses in Improved Network Optimization Algorithms. In *Annual Symposium on Foundations of Computer Science (FOCS 1984)*, 1984.

[60] Daniel Gajski and Longanath Ramachandran. Introduction to High-Level Synthesis. *IEEE Design and Test of Computers*, 2(4):44–54, 1994.

[61] Daniel D. Gajski, Jianwen Zhu, and Rainer Dömer. Essential Issues in Codesign. Technical report, University of California, Irvine, Department of Information and Computer Science, June 1997.

[62] M. R. Garey and D. S. Johnson. *Computers and Intractability: A Guide to the theory of NP-Completeness*. W.H. Freeman and Company, 1979.

[63] Sabih H. Gerez. *Algorithms for VLSI Design Automation*. John Wiley & Sons Ltd., 1998.

[64] David E. Goldberg. *Genetic Algorithms in Search, Optimization & Machine Learning*. Addison-Wesley Publishing Company, 1989.

[65] James Goodman, Anantha Chandrakasan, and Abram P. Dancy. Design and Implementation of a Scalable Encryption Procesoor with Embedded Variable DC/DC Converter. In *Proceedings IEEE 36th Design Automation Conference (DAC99)*, pages 855–860, 1999.

[66] Martin Grajcar. Genetic List Scheduling Algorithm for Scheduling and Allocation on a Loosely Coupled Heterogeneous Multiprocessor System. In *Proceedings IEEE 36th Design Automation Conference (DAC99)*, pages 280–285, 1999.

[67] Flavius Gruian. System-Level Design Methods for Low-Energy Architectures Containing Variable Voltage Processors. In *Workshop Power-Aware Computing Systems*, Nov 2000.

[68] Flavius Gruian and Krzysztof Kuchcinski. LEneS: Task Scheduling for Low-Energy Systems Using Variable Supply Voltage Processors. In *Proceedings Asia South Pacific - Design Automation Conference (ASP-DAC'01)*, pages 449–455, Jan 2001.

[69] R. K. Gupta and G. De Micheli. Hardware/Software Co-synthesis of Digital Systems. *IEEE Design & Test of Computers*, pages 29–41, September 1993.

[70] R.K. Gupta. *Co-Synthesis of Hardware and Software for Digital Embedded Systems*. PhD thesis, Stanford University, December 1993.

[71] Vadim Gutnik and Anantha Chandrakasan. Embedded Power Supply for Low-Power DSP. *IEEE Transactions on VLSI Systems*, 5(4), 425–435 1997.

[72] Johan Hagman. mpeg3play-0.9.6.
Source code available at http://home.swipnet.se/~w-10694/tars/mpeg3play-0.9.6-x86.tar.gz.

[73] Lajos Hanzo, Clare Somerville, and Jason Woodard. *Voice Compression and Communications: Principles and Applications for Fixed and Wireless Channels.* John Wiley & Sons Inc., 2001.

[74] Jörg Henkel. A Low Power Hardware/Software Partitioning Approach for Core-Based Embedded Systems. In *Proceedings IEEE 36th Design Automation Conference (DAC99)*, pages 122–127, 1999.

[75] Jörg Henkel and Rolf Ernst. An Approach to Automated Hardware/Software Partitioning using a Flexible Granularity that is driven by High-Level Estimation Techniques. *IEEE Transactions on VLSI Systems*, 9(2):273–289, 2001.

[76] Inki Hong, Darko Kirovski, Gang Qu, Miodrag Potkonjak, and Mani B. Srivastava. Power Optimization of Variable-Voltage Core-Based Systems. *IEEE Transactions on Computer-Aided Design*, 18(12):1702–1714, Dec 1999.

[77] Inki Hong, Gang Qu, Miodrag Potkonjak, and Mani B. Srivastava. Synthesis Techniques for Low-Power Hard Real-Time Systems on Variable Voltage Processors. In *Proceedings Real-Time Systems Symposium*, 1998.

[78] Junwei Hou and Wayne Wolf. Process Partitioning for Distributed Embedded Systems. In *Proceedings 4th International Workshop Hardware/Software Co-Design (Codes/CASHE'96)*, pages 70 – 76, March 1996.

[79] Sandy Irani, Sandeep Shukla, and Rajesh Gupta. Online Strategies for Dynamic Power Management in Systems with Multiple Power-Saving States. *ACM Transactions on Embedded Computing Systems*, 2(3):325–346, August 2003.

[80] Tohru Ishihara and Hiroto Yasuura. Voltage Scheduling Problem for Dynamically Variable Voltage Processors. In *Proceedings International Symposium Low Power Electronics and Design (ISLPED'98)*, pages 197–202, 1998.

[81] Axel Jantsch and Hannu Tenhunen (Eds.). *Networks on Chip.* Kluwer Academic Publishers, 2003.

[82] Niraj K. Jha. Low Power System Scheduling and Synthesis. In *Proceedings IEEE/ACM International Conference Computer-Aided Design (ICCAD-01)*, pages 259–263, 2001.

[83] Asawaree Kalavade and Edward A. Lee. A Global Criticality/Local Phase Driven Algorithm for the Constrained Hardware/Software Partitioning Problem. In *Proceedings International Workshop Hardware/Software Co-Design (Codes/CASHE'94)*, pages 42–48, Sept. 1994.

[84] Asawaree Kalavade and P. A. Subrahmanyam. Hardware/Software Partitiong for Multifunction Systems. *IEEE Transactions on Computer-Aided Design*, 17(9):819–836, Sep 1998.

[85] Kurt Keutzer, Sharad Malik, A. Richard Newton, Jan M. Rabaey, and A. Sangiovanni-Vincentelli. System-Level Design: Orthogonalization of Concerns and Platform-Based

Design. *IEEE Transactions on Computer-Aided Design*, 19(12):1523–1543, December 2000.

[86] C. Kim and K. Roy. Dynamic Vth Scaling Scheme for Active Leakage Power Reduction. In *Proceedings Design, Automation and Test in Europe Conference (DATE2002)*, pages 163–167, March 2002.

[87] Alexander Klaiber. The Technology behind Crusoe Processors. January 2000. http://www.transmeta.com.

[88] P. V. Knudsen and J. Madsen. Integrating Communication Protocol Selection with Hardware/Software Codesign. *IEEE Transactions on Computer-Aided Design*, 18(9):1077–1095, Aug 1999.

[89] Peter V. Knudsen and Jan Madsen. PACE: A Dynamic Programming Algorithm for Hardware/Software Partitioning. In *Proceedings 4th International Workshop Hardware/Software Co-Design (Codes/CASHE'96)*, pages 85 – 92, March 1996.

[90] Yu-Kwong Kwok and Ishfaq Ahmad. Dynamic Critical-Path Scheduling: An Effective Technique for Allocating Task Graphs to Multiprocessors. *IEEE Transactions on Parallel and Distributed Systems*, 7(5):506–521, May 1996.

[91] Yu-Kwong Kwok and Ishfaq Ahmad. Static Scheduling Algorithms for Allocating Directed Task Graphs to Multiprocessors. *ACM Computing Surveys*, 31(4):406–471, December 1999.

[92] Kanishka Lahiri, Anand Raghunathan, Sujit Dey, and Debashis Panigrahi. Battery-Driven System Design: A New Frontier in Low Power Design. pages ??–??

[93] Rainer Leupers and Peter Marwedel. *Retargetable Compiler Technology for Embedded Systems - Tools and Applications*. Kluwer Academic Publishers, 2001.

[94] Y. Li and J. Henkel. A Framework for Estimating and Minimizing Energy Dissipation of Embedded HW/SW Systems. In *Proceedings IEEE 35th Design Automation Conference (DAC98)*, pages 188–193, 1998.

[95] Y.-T. S. Li, S. Malik, and A. Wolfe. Performance Estimation of Embedded Software with Instruction Cache Modeling. In *Proceedings IEEE/ACM International Conference Computer-Aided Design (ICCAD-95)*, pages 380–387, November 1995.

[96] Jinfeng Liu, Pai H. Chou, and Nader Bagherzadeh. Communication Speed Selection for Embedded Systems with Networked Voltage-Scalable Processors. In *Proceedings 2nd International Symposium Hardware/Software Co-Design (CODES'02)*, pages 169–174, 2002.

[97] J. R. Lorch and A. J. Smith. Software Strategies for Portable Computer Energy Management. *IEEE Personal Communications*, 5(3):60–73, June 1998.

[98] Yung-Hsiang Lu, Luca Benini, and Giovanni De Micheli. Low-Power Task Scheduling for Muliple Devices. In *Proceedings 8th International Workshop Hardware/Software Co-Design (CODES'00)*, pages 39–43, 2000.

[99] Jiong Luo and Niraj K. Jha. Power-conscious Joint Scheduling of Periodic Task Graphs and Aperiodic Tasks in Distributed Real-time Embedded Systems. In *Proceedings*

IEEE/ACM International Conference Computer-Aided Design (ICCAD-00), pages 357–364, Nov 2000.

[100] Jiong Luo and Niraj K. Jha. Battery-aware Static Scheduling for Distributed Real-Time Embedded Systems. In *Proceedings IEEE 38th Design Automation Conference (DAC01)*, pages 444–449, 2001.

[101] Jan Madsen and Peter Bjørn-Jørgensen. Embedded System Synthesis under Memory Constrains. In *Proceedings 7th International Workshop Hardware/Software Co-Design (CODES'99)*, pages 188 – 192, 1999.

[102] S. Martin, K. Flautner, T. Mudge, and D. Blaauw. Combined Dynamic Voltage Scaling and Adaptive Body Biasing for Lower Power Microprocessors under Dynamic Workloads. In *Proceedings IEEE/ACM International Conference Computer-Aided Design (ICCAD-02)*, pages 721–725, 2002.

[103] Sven Mattisson. Minimizing Power Dissipation of Cellular Phones. In *Proceedings International Symposium Low Power Electronics and Design (ISLPED'97)*, pages 42–45, 1997.

[104] G. De Micheli and R. K. Gupta. Hardware/Software Co-Design. In *Proceedings of the IEEE*, pages 349–365, March 1997.

[105] Giovanni De Micheli, Rolf Ernst, and Wayne Wolf. *Readings in Hardware/Software Co-Design*. Morgan Kaufmann Publishers, 2002.

[106] B. Mochocki, X. Hu, and G. Quan. A Realistic Variable Voltage Scheduling Model for Real-Time Applications. In *Proceedings IEEE/ACM International Conference Computer-Aided Design (ICCAD-02)*, pages 726–731, 2002.

[107] J. Monteiro, S. Devadas, and A. Ghosh. Retiming Sequential Circuits for Low Power. In *Proceedings IEEE/ACM International Conference Computer-Aided Design (ICCAD-93)*, pages 398–402, 1993.

[108] J. Monteiro, S. Devadas, and A. Ghosh. Sequential logic optimization for low power using input disabling precomputation architectures. *IEEE Transactions on Computer-Aided Design*, 17(3):279–284, March 1998.

[109] W. Namgoong, M. Yu, and T. Meng. A High-Efficiency Variable-Voltage CMOS Dynamic dc-dc Switching Regulator. In *Proceedings International Solid-State Circuits Conference*, pages 380–381, 1997.

[110] David Naylor and Simon Jones. *VHDL: A Logic Synthesis Approach*. Chapman-Hall, 1997.

[111] Hyunok Oh and Soonhoi Ha. A Static Scheduling Heuristic for Heterogeneous Processors. In *2nd International EuroPar Conference Vol. II*, August 1996.

[112] Hyunok Oh and Soonhoi Ha. A Hardware-Software Cosynthesis Technique Based on Heterogeneous Multiprocessor Scheduling. In *Proceedings 7th International Workshop Hardware/Software Co-Design (CODES'99)*, pages 183–187, May 1999.

[113] Hyunok Oh and Soonhoi Ha. Hardware-Software Cosynthesis of Multi-Mode Multi-Task Embedded Systems with Real-Time Constraints. In *Proceedings 2nd International Symposium Hardware/Software Co-Design (CODES'02)*, pages 133–138, May 2002.

[114] P. R. Panda and N. D. Dutt. Reducing Address Bus Transitions for Low Power Memory Mapping. In *IEEE European Design and Test Conference*, pages 63–67, March 1996.

[115] M. Pedram and Q. Wu. Design Considerations for Battery-powered Electronics. In *Proceedings IEEE 36th Design Automation Conference (DAC99)*, pages 861–866, 1999.

[116] Massoud Pedram. Power Minimization in IC Design: Principles and Applications. *ACM Transactions Design Automation of Electronic Systems (TODAES)*, 1(1):3–56, Jan 1996.

[117] P. Pedro and Alan Burns. Schedulability Analysis for Mode Changes in Flexible Real-time Systems. In *Proceddings Euromicro Workshop on Real-Time Systems*, pages 17–19, June 1998.

[118] Paul Pop, Petru Eles, and Zebo Peng. Bus Access Optimization for Distributed Embedded Systems Based on Schedulability Analysis. In *Proceedings Design, Automation and Test in Europe Conference (DATE2000)*, pages 567–574, 2000.

[119] Paul Pop, Petru Elcs, and Zebo Peng. Scheduling with Optimized Communication for Time-Triggered Embedded Systems. In *Proceedings 8th International Workshop Hardware/Software Co-Design (CODES'00)*, pages 62–66, 2000.

[120] Paul Pop, Petru Eles, Traian Pop, and Zebo Peng. An Approach to Incremental Design of Distributed Embedded Systems. In *Proceedings IEEE 38th Design Automation Conference (DAC01)*, pages 450–455, 2001.

[121] S. Prakash and A. Parker. SOS: Synthesis of Application-Specific Heterogeneous Multiprocessor Systems. *J. Parallel & Distributed Computing*, pages 338–351, Dec 1992.

[122] A. Raghunathan and N.K. Jha. SCALP: An iterative improvement based low-power data path synthesis algorithm. *IEEE Transactions on Computer-Aided Design*, 16(11):1260–1277, November 1997.

[123] A. Raghunathan, N. K. Niraj, and Sujit Dey. *High-Level Power Analysis and Optimization*. Kluwer Academic Publishers, 1998.

[124] Daler Rakhmatov and Sarma Vrudhula. Energy Management for Battery-Powered Embedded Systems. *ACM Transactions on Embedded Computing Systems*, 2(3):277–324, August 2003.

[125] Krithi Ramamritham and John A. Stankovic. Scheduling Algorithms and Operating Systems Support for Real-time Systems. *Proceedings of the IEEE*, 81(1):55–67, 1994.

[126] R. L. Rhodes and Wayne Wolf. Co-Synthesis of Heterogeneous Multiprocessor Systems using Arbitrated Communication. In *Proceedings IEEE/ACM International Conference Computer-Aided Design (ICCAD-99)*, pages 339–342, 1999.

[127] *International Technology Roadmap for Semiconductors*. http://notes.sematech.org/ntrs/PublNTRS.nsf.

[128] Alex Rogers and Adam Prügel-Bennett. Modelling the dynamics of a steady-state genetic algorithm. In *Foundations of Genetic Algorithms (FOGA-5)*, pages 57–68. Sept 1999.

[129] James A. Rowson. Hardware/Software Co-Simulation. In *Proceedings IEEE 31st Design Automation Conference (DAC94)*, pages 439–440, 1994.

[130] Kaushik Roy and Sharat C. Prasad. *Low-Power CMOS VLSI Circuit Design*. Wiley-Interscience, 2000.

[131] Andrew Rushton. *VHDL for Logic Synthesis*. John Wiley & Son Ltd, 1998.

[132] D. Saha, R. S. Mitra, and A. Basu. Hardware Software Partitioning using Genetic Algorithm. In *10th International Conference on VLSI Design*, pages 155–160, January 1997.

[133] Marcus T. Schmitz and Bashir M. Al-Hashimi. Considering Power Variations of DVS Processing Elements for Energy Minimisation in Distributed Systems. In *Proceedings International Symposium System Synthesis (ISSS'01)*, pages 250–255, October 2001.

[134] E. M. Sentovitch, K. J. Singh, Luciano Lavagno, Cho Moon, Rajeev Murgai, Alexander Saldanha, Hamid Savoj, Paul R. Stephan, Robert K. Brayton, and Alberto Sangiovanni-Vincentelli. SIS: A System for Sequential circuit synthesis. Technical report, University of California, Berkeley, May 1992.

[135] L. Sha, R. Rajkumar, J. Lehoczky, and K Ramamritham. Mode Change Protocols for Priority-driven Preemptive Scheduling. 1:243–265, December 1989.

[136] Y. Shin, D. Kim, and K. Choi. Schedulability-Driven Performance Analysis of Multiple Mode Embedded Real-Time Systems. In *Proceedings IEEE 37th Design Automation Conference (DAC00)*, pages 495–500, June 2000.

[137] Youngsoo Shin and Kiyoung Choi. Power Conscious Fixed Priority Scheduling for Hard Real-Time Systems. In *Proceedings IEEE 36th Design Automation Conference (DAC99)*, pages 134–139, 1999.

[138] Gilbert C. Sih and Edward A. Lee. A Compile-time scheduling heuristic for interconnection-constrained heterogeneous processor architectures. *IEEE Transactions on Parallel and Distributed Systems*, 4(2):175–187, February 1993.

[139] T. Simunic, L. Benini, and G. De Micheli. Energy-Efficient Design of Battery-Powered Embedded Systems. In *Proceedings International Symposium Low Power Electronics and Design (ISLPED'99)*, pages 212–217, 1999.

[140] M. Srivastava, A. Chandrakasan, and R. Brodersen. Predictive System Shutdown and other Architectural Techniques for Energy Efficient Programmable Computations. *IEEE Transactions on VLSI Systems*, 4(1):42–55, March 1996.

[141] M. R. Stan and W. P. Burleson. Bus-Inverse Coding for Low-Power I/O. *IEEE Transactions on VLSI Systems*, 3(1):49–58, March 1995.

[142] Jorgen Staunstrup and Wayne Wolf. *Hardware/Software Co-Design: Principles and Practice*. Kluwer Academic Publishers, 1997.

[143] J. Teich, T. Blickle, and Lothar Thiele. An Evolutionary Approach to System-Level Synthesis. In *Proceedings 5th International Workshop Hardware/Software Co-Design (Codes/CASHE'97)*, pages 167 – 171, March 1997.

[144] V. Tiwari and M. Lee. Power Analysis of a 32-bit Embedded Microcontroller. In *ASP-DAC*, pages 141–148, Aug 1995.

[145] V. Tiwari, S. Malik, A. Wolfe, and M. Lee. Instruction Level Power Analysis and Optimization of Software. *Journal of VLSI Signal Processing Systems*, 13(2–3):223–238, 1996.

[146] Vivek Tiwari, Sharad Malik, and Andrew Wolfe. Power Analysis of Embedded Software: A First Step Towards Software Power Minimization. *IEEE Transactions on VLSI Systems*, Dec 1994.

[147] C-Y. Tsui, J. Monteiro, M. Pedram, S. Devadas, A. Despain, and B. Lin. Precomputation-Based Sequential Logic Optimization for Low Power. *IEEE Transactions on VLSI Systems*, 2(4):426–436, December 1994.

[148] K. S. Vallerio and N. K. Jha. Task graph extraction for embedded system synthesis. In *International Conference on VLSI Design*, pages 480–486, January 2003.

[149] Bhaskaran Vasuder. *Image and Video Compression Standards*. Kluwer Academic Publisher, 1997.

[150] Matthew Wall. GAlib: A C++ Library of Genetic Algorithm Components Version 2.45, August 1996. available at: http://lancet.mit.edu/ga.

[151] John Watkinson. *The engineer's guide to compression*. Snell & Wilcox, 1996.

[152] M. Weiser, B. Welch, A. Demers, and S. Shenker. Scheduling for Reduced CPU Energy. In *Proceedings USENIX Symposium on Operating Systems Design and Implementation (OSDI)*, pages 13–23, 1994.

[153] T. Wiangtong, Peter Y.K. Cheung, and W. Luk. Comparing Three Heuristic Search Methods for Functional Partitioning in Hardware-Software Codesign. *Design Automation for Embedded Systems*, 6(4):425–449, July 2002.

[154] A. C. Williams. *A Behavioural VHDL Synthesis System using Data Path Optimisation*. PhD thesis, University of Southampton, October 1997.

[155] A. C. Williams, A. D. Brown, and M. Zwolinski. Simultaneous optimisation of dynamic power, area and delay in behavioural synthesis. *IEE Proc.-Comput. Digit. Tech.*, 147(6):383–390, November 2000.

[156] Wayne H. Wolf. Hardware/Software Co-Design of Embedded Systems. In *Proceedings of the IEEE*, pages 967–989, July 1994.

[157] Wayne H. Wolf. An Architectural Co-Synthesis Algorithm for Distributed, Embedded Computing Systems. *IEEE Transactions on VLSI Systems*, 5(2):218–229, June 1997.

[158] M. Wu and D. Gajski. Hypertool: A Programming Aid for Message-passing Systems. *IEEE Transactions on Parallel and Distributed Systems*, 1(3):330–343, July 1990.

[159] Y. Xie and Wayne Wolf. Allocation and Scheduling of Conditional Task Graph in Hardware/Software Co-Synthesis. In *Proceedings Design, Automation and Test in Europe Conference (DATE2001)*, pages 620 – 625, March 2001.

[160] Peng Yang, Paul Marchal, Chun Wong, Stefaan Himpe, Francky Catthoor, Patrick David, Johan Vounckx, and Rudy Lauwereins. Managing Dynamic Concurrent Tasks in Embedded Real-Time Multimedia Systems. In *Proceedings International Symposium System Synthesis (ISSS'02)*, pages 112–119, October 2002.

[161] Frances Yao, Alan Demers, and Scott Shenker. A Scheduling Model for Reduced CPU Energy. In *IEEE Symposium on Foundations of Comp. Science*, pages 374–382, 1995.

[162] Y. Zhang, X. Hu, and D. Chen. Energy Minimization of Real-time Tasks on Variable Voltage Processors with Transition Energy Overhead. In *Proc. ASP-DAC'03*, pages 65–70, 2003.

[163] Yumin Zhang, Xiaobo (Sharon) Hu, and Denny Z. Chen. Task Scheduling and Voltage Selection for Energy Minimization. In *Proceedings IEEE 39th Design Automation Conference (DAC02)*, pages 183–188, 2002.

Index